AF332928

MARINE BIOLOGY

AQUATIC INVASIVE SPECIES

FEDERAL ACTIVITIES AND COSTS OF ADDRESSING THREATS AND IMPACTS

MARINE BIOLOGY

Additional books in this series can be found on Nova's website
under the Series tab.

Additional e-books in this series can be found on Nova's website
under the eBooks tab.

MARINE BIOLOGY

AQUATIC INVASIVE SPECIES

FEDERAL ACTIVITIES AND COSTS OF ADDRESSING THREATS AND IMPACTS

ERNESTINE SANDOVAL
EDITOR

New York

Copyright © 2016 by Nova Science Publishers, Inc.

All rights reserved. No part of this book may be reproduced, stored in a retrieval system or transmitted in any form or by any means: electronic, electrostatic, magnetic, tape, mechanical photocopying, recording or otherwise without the written permission of the Publisher.

We have partnered with Copyright Clearance Center to make it easy for you to obtain permissions to reuse content from this publication. Simply navigate to this publication's page on Nova's website and locate the "Get Permission" button below the title description. This button is linked directly to the title's permission page on copyright.com. Alternatively, you can visit copyright.com and search by title, ISBN, or ISSN.

For further questions about using the service on copyright.com, please contact:
Copyright Clearance Center
Phone: +1-(978) 750-8400   Fax: +1-(978) 750-4470   E-mail: info@copyright.com

NOTICE TO THE READER

The Publisher has taken reasonable care in the preparation of this book, but makes no expressed or implied warranty of any kind and assumes no responsibility for any errors or omissions. No liability is assumed for incidental or consequential damages in connection with or arising out of information contained in this book. The Publisher shall not be liable for any special, consequential, or exemplary damages resulting, in whole or in part, from the readers' use of, or reliance upon, this material. Any parts of this book based on government reports are so indicated and copyright is claimed for those parts to the extent applicable to compilations of such works.

Independent verification should be sought for any data, advice or recommendations contained in this book. In addition, no responsibility is assumed by the publisher for any injury and/or damage to persons or property arising from any methods, products, instructions, ideas or otherwise contained in this publication.

This publication is designed to provide accurate and authoritative information with regard to the subject matter covered herein. It is sold with the clear understanding that the Publisher is not engaged in rendering legal or any other professional services. If legal or any other expert assistance is required, the services of a competent person should be sought. FROM A DECLARATION OF PARTICIPANTS JOINTLY ADOPTED BY A COMMITTEE OF THE AMERICAN BAR ASSOCIATION AND A COMMITTEE OF PUBLISHERS.

Additional color graphics may be available in the e-book version of this book.

Library of Congress Cataloging-in-Publication Data

ISBN: 978-1-63485-391-0

Published by Nova Science Publishers, Inc. † New York

CONTENTS

PREFACE

Aquatic invasive species—harmful, nonnative plants, animals, and microorganisms living in aquatic habitats—damage ecosystems or threaten commercial, agricultural, and recreational activities. The Nonindigenous Aquatic Nuisance Prevention and Control Act of 1990 created the Task Force and required it to develop an aquatic nuisance species program. The Water Resources Reform and Development Act of 2014 includes a provision that the United States Government Accountability Office (GAO) assess federal costs of, and spending on, aquatic invasive species. This book examines how much Task Force member agencies expended addressing aquatic invasive species for fiscal years 2012-2014; activities conducted by Task Force member agencies and challenges in addressing aquatic invasive species; and the extent to which the Task Force has measured progress in achieving the goals of its 2013-2017 strategic plan.

In: Aquatic Invasive Species
Editor: Ernestine Sandoval

ISBN: 978-1-63485-391-0
© 2016 Nova Science Publishers, Inc.

Chapter 1

AQUATIC INVASIVE SPECIES: ADDITIONAL STEPS COULD HELP MEASURE FEDERAL PROGRESS IN ACHIEVING STRATEGIC GOALS[*]

United States Government Accountability Office

WHY GAO DID THIS STUDY

Aquatic invasive species—harmful, nonnative plants, animals, and microorganisms living in aquatic habitats—damage ecosystems or threaten commercial, agricultural, and recreational activities. The Nonindigenous Aquatic Nuisance Prevention and Control Act of 1990 created the Task Force and required it to develop an aquatic nuisance (which GAO refers to as invasive) species program. The Water Resources Reform and Development Act of 2014 includes a provision that GAO assess federal costs of, and spending on, aquatic invasive species.

This report examines (1) how much Task Force member agencies expended addressing aquatic invasive species for fiscal years 2012-2014; (2) activities conducted by Task Force member agencies and challenges in addressing aquatic invasive species; and (3) the extent to which the Task Force has measured progress in achieving the goals of its 2013-2017 strategic plan.

---

[*] This is an edited, reformatted and augmented version of a United States Government Accountability Office, Publication No. GAO-16-49, dated November 2015.

GAO sent a questionnaire to member agencies to obtain expenditures for fiscal years 2012-2014; interviewed member agency officials; and analyzed laws and strategic planning documents.

WHAT GAO RECOMMENDS

GAO recommends that the Task Force develop a mechanism to measure progress toward its strategic goals and help meet certain statutory requirements. Most member agencies generally concurred or had no comments, but NOAA disagreed. GAO believes its recommendation is valid as discussed further in this report.

WHAT GAO FOUND

The 13 federal member agencies of the Aquatic Nuisance Species Task Force (Task Force) estimated expending an average of about $260 million annually for fiscal years 2012 through 2014 to address aquatic invasive species. However, several member agencies identified in their questionnaire responses challenges in developing their estimates. For example, some member agencies reported that their activities to address aquatic invasive species were often integrated into larger projects, making it difficult to isolate the portion of expenditures specific to aquatic invasive species out of total expenditures for the projects. As a result, expenditure information reported by GAO generally reflects member agencies' best estimates of total expenditures, rather than actual expenditures.

Task Force member agencies conducted a wide range of activities and identified several challenges in addressing aquatic invasive species. Member agencies reported conducting activities across several activity categories, including taking actions to prevent introductions, control the spread of existing invaders, and research ecological impacts of aquatic invasive species. For instance, most conducted prevention activities—such as constructing a series of electric barriers to prevent the entry of Asian Carp from the Mississippi River Basin into the Great Lakes—recognizing that prevention activities may be the most cost-effective method of addressing aquatic invasive species. Additionally, officials from several member agencies expressed concern that their activities, though numerous, may not be adequate relative to the growing magnitude and impacts of aquatic invasive species amid decreasing or constrained agency resources.

The Task Force—which is co-chaired by the U.S. Fish and Wildlife Service and National Oceanic and Atmospheric Administration (NOAA)—developed a 2013- 2017 strategic plan to guide its member agencies but has not taken key steps to measure progress in achieving the goals laid out in its strategic plan. As called for in its strategic plan, the Task Force in 2012 planned to develop an operational plan to track and measure aquatic invasive species activities and progress. However, the Task Force did not develop an operational plan because of constrained funding and limited resources, according to Task Force representatives. The Task Force also did not meet several of the 1990 Act's requirements including describing its members' roles and activities and reporting annually to Congress on the program's progress. The representatives agreed that a mechanism to track activities and measure progress is important and said they plan to discuss the possibility of doing so at their November 2015 meeting. Task Force representatives, however, had not established a time frame or specifics for their approach. Developing and regularly using a tracking mechanism could help the Task Force measure progress in achieving its strategic goals, as well as help the Task Force meet the 1990 Act's requirements to describe its members' roles and specific activities and to report annually to Congress on the program's progress.

ABBREVIATIONS

Corps	U.S. Army Corps of Engineers
EPA	Environmental Protection Agency
FWS	U.S. Fish and Wildlife Service
NOAA	National Oceanic and Atmospheric Administration
Task Force	Aquatic Nuisance Species Task Force
The 1990 Act	The Nonindigenous Aquatic Nuisance Prevention and Control Act of 1990
USGS	U.S. Geological Survey

November 30, 2015

Congressional Committees

Invasive species[1]—harmful, nonnative plants, animals, and microorganisms—are pervasive throughout the United States and cause major economic losses to segments of the economy and significant environmental damage each year to crops, rangelands, waterways, fisheries, and ecosystems.[2] Invasive species, called a national crisis by the Department of the Interior,[3] number in the thousands and are expected to increase, with about 250 new species having invaded the United States since 2011 and more than 750 invasive species expanding their range since that time.[4] As we have found, the impact of invasive species in the United States is widespread, and their consequences for the economy and the environment are profound, although this can be difficult to measure.[5] A widely cited academic study from 2005—the most recent comprehensive study of its kind—estimated that the environmental impacts and economic costs associated with invasive species amount to almost $120 billion per year.[6]

Addressing aquatic invasive species is a complex, interdisciplinary issue with the potential to affect many sectors and levels of government operations. Multiple federal agencies, often in coordination with state and local governments, industry, international parties, and nongovernmental agencies, work to prevent, manage, eradicate, and raise awareness about invasive species. The National Invasive Species Council, which was established by an Executive Order in 1999 to, among other things, coordinate federal agencies' activities concerning invasive species,[7] reported that estimated expenditures for invasive species activities by more than 20 federal agencies were over $2 billion dollars in fiscal year 2014.[8] This estimate encompasses expenditures for all invasive species, however, and does not separate out expenditures specific to aquatic invasive species—species found in marine, freshwater, estuarine, and riparian areas, such as fish, mollusks, snakes, plants, and pathogens or parasites of aquatic animals and plants. Aquatic invasive species, which are one type of invasive species, harm native ecosystems or commercial, agricultural, or recreational activities dependent on these ecosystems, such as by threatening commercially or recreationally important fish species, according to the National Invasive Species Council. Officials from National Oceanic and Atmospheric Administration (NOAA) and the Department of the Interior have likened aquatic invasive species to an oil spill that will continue to spread unless promptly and completely contained—once they have arrived and become established, aquatic invasive species are difficult to eradicate.[9]

The Nonindigenous Aquatic Nuisance Prevention and Control Act of 1990, as amended (the 1990 Act), was enacted to, among other things, prevent the unintentional introduction and dispersal of nonindigenous (which we refer to as nonnative) species into waters of the United States through the management of ballast water (water in a ship's holding tank used for stability and safety that may be taken on in one location and discharged in another) and other requirements, and to understand and minimize economic and ecological impacts of nonnative aquatic nuisance species that become established in the United States.[10] The 1990 Act notes that, if preventive management measures are not taken nationwide to prevent and control unintentionally introduced nonnative aquatic species in a timely manner, further introductions and infestations of destructive species may occur. The 1990 Act created the Aquatic Nuisance Species Task Force (Task Force), which coordinates governmental efforts dealing with aquatic invasive species in the United States through regional panels, special committees, and work groups. The Task Force consists of 13 federal member agencies along with state, regional, and nongovernmental organizations.[11] Each of the Task Force's federal member agencies has a different set of responsibilities related to aquatic invasive species.[12]

The 1990 Act requires, among other things, the Task Force to develop and implement an aquatic invasive species program for waters of the United States. Specifically, the 1990 Act requires the Task Force to develop a program that identifies the goals, priorities, and approaches for aquatic invasive species prevention, monitoring, control, education, and research to be conducted or funded by the federal government. The act requires the Task Force to (1) describe the specific prevention, monitoring, control, education, and research activities to be conducted by each Task Force member; (2) describe the role of each Task Force member in implementing the elements of the program; and (3) include recommendations for funding to implement elements of the program.[13] The act also requires that the Task Force report to Congress annually on the progress of its program.[14] In 1994, The Task Force developed a program overview that established the core elements of its aquatic invasive species program and served to guide the work of the Task Force. The Task Force subsequently developed a series of strategic plans starting in the early 2000s to further guide its membership—the federal, state, regional, and nongovernmental organizations that conduct aquatic invasive species activities—in implementing the aquatic invasive species program. In 2012, the

Task Force developed its most recent strategic plan, covering 2013 through 2017, which identified eight goals for the program.

The Water Resources Reform and Development Act of 2014 includes a provision in section 1039(a)(2) that GAO conduct an assessment of the federal costs of, and spending on, aquatic invasive species.[15] We briefed your offices on our preliminary results on June 3, 2015. This report transmits our final results related to this review. This report examines (1) how much Task Force member agencies expended addressing aquatic invasive species for fiscal years 2012 through 2014; (2) activities conducted by Task Force member agencies and challenges in addressing aquatic invasive species; and (3) the extent to which the Task Force has measured progress in achieving the goals of its 2013-2017 strategic plan.

For all three objectives, we reviewed aquatic invasive species-related laws, regulations, and academic studies. To determine how much Task Force member agencies expended addressing aquatic invasive species and to obtain information on activities conducted, we conducted interviews with, and obtained documentation from the Task Force and its 13 federal member departments and agencies (member agencies) regarding any expenditure information they maintain related to aquatic invasive species. We also interviewed staff from the National Invasive Species Council to learn about their efforts to collect information on federal expenditures for invasive species activities. We then developed and disseminated a questionnaire to the 13 Task Force member agencies to obtain each member agencies' estimated annual expenditures to address aquatic invasive species for fiscal years 2012 through 2014 (the most recent years for which reliable data were available) and examples of these activities.[16] The expenditures reflect the agencies' best estimates of how much they spent on aquatic invasive species activities during these years. Based on our assessment of the estimated annual expenditures reported by Task Force member agencies, we found the estimates for fiscal years 2012 through 2014 were sufficiently reliable for purposes of this report—to provide general estimates of total annual expenditures by these agencies on activities related to aquatic invasive species.

To further describe activities conducted by Task Force member agencies and any challenges in addressing aquatic invasive species, we built on the information gathered through our questionnaire and conducted a series of interviews with officials from the 13 member agencies; the federal ex-officio member of the Task Force, the Smithsonian Environmental Research Center; and each of the Task Force's six regional panels. Through these interviews, we collected information and documentation on the agencies' aquatic invasive species activities and any challenges they face in addressing aquatic invasive species. Many of the activities reported by agencies were ongoing or span multiple fiscal years, and thus, the information we collected often highlights, but is not limited to, fiscal years 2012 through 2014. We also conducted site visits in Southern Florida, Northern California, and Western Washington to observe activities and interview local federal officials at the sites. We selected these locations based on the number and variety of aquatic invasive species present and federal agencies involved, as well as the types of activities conducted in those locations.

To determine the extent to which the Task Force has measured progress in achieving the goals of its 2013-2017 strategic plan, we conducted interviews with and obtained documentation from Task Force representatives, officials from the 13 Task Force member agencies, and officials representing the six regional panels. We reviewed the Task Force's 2013-2017 strategic plan and other documentation related to its strategic plan. We then compared this information to program requirements identified in the 1990 Act, our previous reports on leading practices provided by the Government Performance and Results Modernization Act of 2010, and our executive guide on strategic planning,[17] as appropriate. Appendix I presents a more detailed description of our objectives, scope, and methodology.

We conducted this performance audit from November 2014 to November 2015 in accordance with generally accepted government auditing standards. Those standards require that we plan and perform the audit to obtain sufficient, appropriate evidence to provide a reasonable basis for our findings and conclusions based on our audit objectives. We believe that the evidence obtained provides a reasonable basis for our findings and conclusions based on our audit objectives.

BACKGROUND

Aquatic invasive species can be found in all U.S. states and territories. They can enter and travel in aquatic habitats by several common pathways, including through the discharge of ships' ballast water; hull fouling, such as barnacle growth, on commercial vessels and recreational boats; and accidental or intentional release of organisms into aquatic habitats through aquaculture, bait, aquaria (fish tanks), or the pet trade. Once established in a particular location, an aquatic invasive species can spread to other locations and ecosystems. Scientists and officials from several federal agencies said that the presence and impacts of aquatic invasive species are, and are likely to continue, growing, such as from the warming of ocean waters and the opening of shipping channels through the Arctic, allowing new species to potentially thrive in habitats previously too cold or inaccessible.

See Appendix II for more information on species examples, their known locations, and invasion pathways.

The Task Force, created by the 1990 Act, is co-chaired by the U.S. Fish and Wildlife Service (FWS) and NOAA. FWS provides funding for the administration of the Task Force, including conducting annual meetings, publishing *Federal Register* notices, and supporting an Executive Secretary and other FWS staff that work as regional coordinators.[18] To implement its aquatic invasive species program, the Task Force relies on its 13 member agencies—each of which has a different set of responsibilities related to aquatic invasive species, based on their overall mission and areas of programmatic responsibility (see table 1). These member agencies conduct aquatic invasive species activities and commit resources to achieve the goals of the aquatic invasive species program.[19] According to the Task Force's 1994 program overview, implementation of the program is a cooperative effort that will build on and fill gaps in existing activities and programs, and individual agencies will implement the program in line with their specific authorities, priorities, expertise, and funding. In addition, the Task Force is advised by six regional panels— consisting of representatives of state, tribal, and nongovernmental organizations, commercial interests, and neighboring countries—that help identify regional priorities and coordinate regional activities.[20] Some funding is provided to each regional panel as well as to state governments and other entities to support implementation of species- or region-specific aquatic invasive species management plans and other activities.[21] Together, these federal, state, and nonfederal agencies and organizations work to prevent and control aquatic invasive species and implement the 1990 Act.

Activities to address aquatic invasive species can be categorized using the seven general activity categories developed by the National Invasive Species Council. These categories reflect common activities agencies conduct along the continuum of an invasion of a species, from preventing the arrival or spread of an invading species to controlling or eradicating that species from the ecosystem. Table 2 describes each activity category.

Table 1. Aquatic Nuisance Species Task Force Member Agencies' Key Roles and Responsibilities

Department or agency	Component agency	Key roles and responsibilities for aquatic invasive species
Department of Agriculture	Animal and Plant Health Inspection Service	Manage the Federal Noxious Weed Lista and inspect imports for prohibited aquatic plants. Control and manage aquatic animals and animal pathogens with significant impact on commercial aquaculture.
	U.S. Forest Service	Control and prevent aquatic invasive species on national forest land and participate in prevention efforts through educational outreach.
Department of Commerce	National Oceanic and Atmospheric Administration	Co-chair the Task Force. Responsible for prevention, monitoring, control, education, and research to prevent introductions and spread of aquatic invasive species in aquatic habitats such as National Marine Sanctuaries.
Department of Defense	U.S. Army Corps of Engineers	Control and manage the spread of aquatic invasive species on or around the facilities, ports, harbors, and other navigable waterways it manages. Support research programs on aquatic invasive species.
Department of Homeland Security	U.S. Coast Guard	Regulate biofouling (the accumulation of organisms on a vessel's exterior surfaces) and ballast water to help ensure aquatic invasive species are not discharged into waters of the United States.
Department of the Interior	Bureau of Land Management	Prevent and control the impacts of aquatic invasive plants and animals on fish and wildlife habitat throughout the public lands it manages.
	Bureau of Reclamation	Manage programs to control aquatic invasive species in water systems it manages, including reservoirs, rivers, and distribution canals.
	National Park Service	Manage and control aquatic invasive species established in national parks and prevent the introduction of new species in national parks.
	U.S. Fish and Wildlife Service	Co-chair the Task Force. Manage multiple programs addressing prevention, management, and control of aquatic invasive species. Enforce the Lacey Act,b including its prohibitions on the importation and interstate transport of injurious wildlife (which includes certain aquatic invasive species).

Table 1. (Continued)

Department or agency	Component agency	Key roles and responsibilities for aquatic invasive species
	U.S. Geological Survey	Conduct research and provide scientific information on aquatic invasive species to support the efforts of federal agencies and other partners.
Department of State		Fund the Great Lakes Fishery Commission and other agencies' efforts to address aquatic invasive species and coordinate with international parties.
Department of Transportation	Maritime Administration	Research ballast water and hull fouling issues and participate in ship hull cleaning efforts for ship disposal.
Environmental Protection Agency		Regulate discharges incidental to the normal operation of certain vessels, such as ballast water, under Sec. 402 of the Clean Water Act. Chair the Great Lakes Interagency Task Force overseeing the Great Lakes Restoration Initiative, which includes efforts to address aquatic invasive species.

Sources: Draft, Aquatic Nuisance Species Task Force 2014 Report to Congress (information); GAO (analysis). | GAO-16-49

Note: This list of agency roles and responsibilities for aquatic invasive species is not comprehensive; it is intended to illustrate key roles and responsibilities of each agency as they relate to aquatic invasive species.

[a]The Plant Protection Act authorizes the Secretary of Agriculture, who has delegated this authority to the Animal and Plant Health Inspection Service to publish, by regulation, a list of noxious weeds and to prohibit or restrict their importation, exportation, or movement in interstate commerce if necessary to prevent the introduction into the United States or the dissemination of a noxious weed within the United States (7 U.S.C. § 7712). Noxious weeds are any plant or plant product that can directly or indirectly injure or cause damage to crops, livestock, poultry, or other interests of agriculture, irrigation, navigation, the natural resources of the United States, the public health, or the environment.

[b]Among other things, the Lacey Act, as amended, prohibits the importation into the United States and interstate transport of certain animals, including aquatic invasive species listed as injurious wildlife.

Table 2. Categories of Activities to Address Invasive Species

Activity category	Definition
Prevention	Actions taken to prevent the entry, establishment, dispersal, and dissemination of invasive species.
Early detection and rapid response	Actions taken to detect incipient invasions and assess the current and potential impact of invasions; to eradicate, contain, or control a potentially invasive nonnative species introduced into an ecosystem while the infestation of that ecosystem is still localized, and to eradicate or contain invasive species populations while they are still localized.
Control and management	Actions taken to lessen and manage the impact of invasive species within their established ranges and limit their spread.
Restoration	Actions taken to assist the recovery and reestablishment of plant and animal communities that have been overwhelmed by invasive species.
Research	Actions taken to identify, evaluate, control, and understand invasive species and their interactions with the biotic (i.e., living things that shape an ecosystem) and abiotic (i.e., not derived from living organisms) elements of the environment.
Education and public awareness	Actions taken to maintain and increase public awareness of invasive species and related programs and to promote public activities that reduce the spread and impact of invasive species.
Leadership and international cooperation	Actions taken to provide leadership, oversight, and coordination to maintain and enhance the capabilities to prevent, control, manage, and understand invasive species and invasion pathways with relevant state, local and international partners, and provide for public input and participation.

Sources: The National Invasive Species Council Invasive Species Interagency Crosscut Budget Summary (2014); GAO questionnaire of 13 Task Force member agencies. | GAO-16-49

Note: These activity categories apply to all invasive species, including both aquatic and terrestrial species.

Preventing the introduction of aquatic invasive species into ecosystems is generally the most effective means of avoiding their establishment and spread, according to numerous academic reports, as well as the Task Force and several of its member agencies.[22] According to a 2006 study, the difficulties and expense of reversing biological invasions means investment in prevention is likely to be the most successful and cost-effective response to biological invasions.[23] Further, eradication (the elimination of an invading species from the ecosystem) and control (limiting an invasive species to a specific ecosystem) becomes increasingly difficult and costly as a species becomes established and spreads, as shown in figure 1.[24]

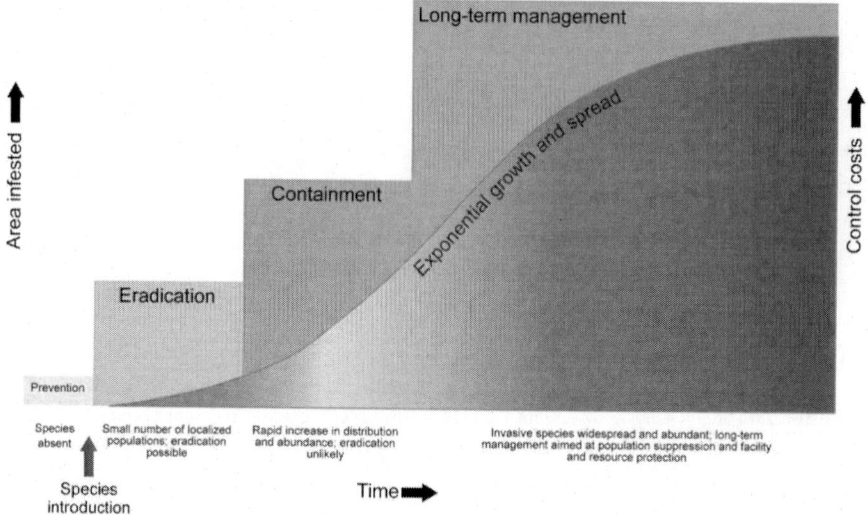

Sources: National Invasive Species Council; U.S. Department of Agriculture; National Park Service, U.S. Fish and Wildlife Service; Rodgers, L., South Florida Water Management District; Department of Primary Industries, State of Victoria; Australia; and GAO. | GAO-16-49.

Figure 1. Invasive Species Invasion Curve.

TASK FORCE MEMBER AGENCIES ESTIMATED EXPENDING AN AVERAGE OF ABOUT $260 MILLION ANNUALLY TO ADDRESS AQUATIC INVASIVE SPECIES IN FISCAL YEARS 2012 THROUGH 2014

Task Force member agencies estimated expending an average of about $260 million annually for fiscal years 2012 through 2014 to address aquatic invasive species. Several of the member agencies identified challenges and limitations associated with the expenditure information they provided in response to our questionnaire. As a result, the information reported by Task Force member agencies on annual expenditures through our questionnaire generally reflects the agencies' best estimates, rather than actual expenditures. Table 3 provides the estimated annual expenditures for each Task Force member agency during fiscal years 2012 through 2014.

Table 3. Estimated Annual Expenditures to Address Aquatic Invasive Species, by Task Force Member Agency, Fiscal Years 2012 through 2014

Dollars in thousands	Fiscal year 2012 estimated expenditures	Fiscal year 2013 estimated expenditures	Fiscal year 2014 estimated expenditures
Department of Agriculture			
Animal and Plant Health Inspection Service	$3,618	$3,454	$3,467
U.S. Forest Service	a	a	a
Department of Commerce			
National Oceanic and Atmospheric Administration	3,855	3,454	3,528
Department of Defense			
U.S. Army Corps of Engineers	134,886	133,139	148,571
Department of Homeland Security			
U.S. Coast Guard	2,809	3,995	3,362
Department of the Interior			
Bureau of Land Management	298	205	70
Bureau of Reclamation	4,728	5,352	6,335
National Park Service	3,967	4,078	5,437
U.S. Fish and Wildlife Service	13,049	13,096	15,049
U.S. Geological Survey	10,229	10,034	11,325
Department of State[b]	18,575	18,106	19,341
Department of Transportation			
Maritime Administration	5,137	4,965	4,473
Environmental Protection Agency[c]	56,709	44,946	54,599
Total	**$257,860**	**$244,824**	**$275,557**

Sources: GAO questionnaires of 13 Task Force member agencies. I GAO-16-49

Note: The data in this table were provided by Task Force member agencies and reflect their best estimates of annual expenditures to address aquatic invasive species for fiscal years 2012 through 2014.

[a] The U.S. Forest Service reported that it was unable to develop estimated annual expenditures for its aquatic invasive species activities. The agency cited several reasons, including that its program management and financial accounting systems do not separately track aquatic invasive species expenditures.

[b] The Department of State estimates reflect amounts mostly provided to the Great Lakes Fishery Commission, largely for Sea Lamprey control and research efforts. The Great Lakes Fishery Commission provided financial resources to other Task Force member agencies and therefore to minimize double-counting, we have to the extent possible, excluded this funding from the estimates reported by the other Task Force member agencies.

[c] The Environmental Protection Agency's estimates reflect amounts for the Great Lakes Restoration Initiative (a program launched in 2010 to protect and restore the Great Lakes ecosystem), most of which was provided to other Task Force member agencies. To minimize double-counting, we have, to the extent possible, excluded this funding from the estimates reported by the other Task Force member agencies.

Based on information reported through our questionnaire, estimated expenditures by Task Force member agencies for fiscal year 2014 ranged from a high of about $149 million by the U.S. Army Corps of Engineers (Corps) to a low of $70,000 by the Bureau of Land Management. Specifically, the Corps reported that the majority of its estimated annual expenditures were for controlling and managing existing aquatic invasive species at multiple projects it manages, and mostly came from the respective project's operations and maintenance funding. The Bureau of Land Management's estimates for fiscal year 2014 comprised the annual cost to develop and place aquatic invasive species awareness advertisements in print materials focusing on outdoor activities such as hunting, fishing, and boating. Also, for fiscal year 2014, the Bureau of Land Management reported that it did not have funding to provide to its state offices to coordinate or carry out aquatic invasive species activities in their local areas, as it did in fiscal years 2012 and 2013. Estimates for Task Force member agencies generally reflected a variety of activities undertaken or funded by the respective agency spanning multiple species and regions within their areas of programmatic responsibility. In contrast, estimates for some agencies reflected efforts specific to a particular region or activity. For example, the Environmental Protection Agency (EPA) reported that its estimates mostly reflected expenditures of funding transferred to other agencies to carry out activities in support of the Great Lakes Restoration Initiative—a program launched in 2010 to protect and restore the Great Lakes ecosystem. One of the Initiative's main focus areas includes prioritizing efforts to prevent the introduction of new invasive species into the Great Lakes.[25]

In responding to our questionnaire, several of the Task Force member agencies identified challenges and limitations in collecting information on how much they estimated expending to address aquatic invasive species. These included the following:

- **Expenditures on aquatic invasive species activities are not specifically tracked.** Seven of the 13 Task Force member agencies reported that their budget structures and financial accounting systems were not designed to specifically track expenditures on aquatic invasive species activities. For instance, the U.S. Forest Service reported that many aquatic invasive species related activities are conducted throughout the agency, but the agency's program management and financial accounting systems do not separately track aquatic invasive species expenditures. Specifically, U.S. Forest Service officials said they could not identify the portion of funding expended directly for aquatic invasive species because these activities were often integrated into larger projects—such as inspecting and cleaning equipment used in fighting wildfires. For example, the agency has developed specific protocols to inspect, assess, and decontaminate equipment, such as the inside of a fire pump, to help make sure it is clear of any invasive algae or mussels that may be unintentionally transferred to a new watershed when moving water between areas to fight fires. U.S. Forest Service officials further explained that this is one step of many in cleaning and preparing the equipment for its next use, and its management and financial accounting systems are not set up to capture or break out activities to this level of detail. Similarly, the Bureau of Reclamation reported that expenditures for aquatic invasive species activities at its water projects—such as clearing water control structures to maintain water delivery through pipes and canals—are funded mostly through the operations and maintenance budget for each project and are not tracked as expenditures specific to aquatic invasive species.

- **Decisions on expenditures for aquatic invasive species are made at the local or regional level.** Four of the 13 member agencies reported that decisions on expenditures for aquatic invasive species activities are delegated to a regional or local level and are not tracked at the national level. For example, the National Park Service reported that once funding is provided to a national park, headquarters management does not generally direct how the funding is expended at that park. Instead, park management generally determines how the funding will be used to accomplish park objectives, including whether and how to prioritize funding for aquatic invasive species activities. Similarly, the Bureau of Land Management reported that numerous decisions and activities take place at its local or state office level that are not tracked by headquarters, including expenditures on aquatic invasive species, and, therefore, annual expenditures on aquatic invasive species across the agency are unknown. The U.S. Forest Service also reported in its questionnaire that many of its aquatic invasive species activities are conducted through cooperative partnership agreements at the local and regional level and expenditures for these activities are not reported at the national level.

TASK FORCE MEMBER AGENCIES CONDUCTED A WIDE RANGE OF ACTIVITIES AND IDENTIFIED SEVERAL CHALLENGES IN ADDRESSING AQUATIC INVASIVE SPECIES

Through our questionnaire and interviews with officials from the Task Force and its member agencies, we found that member agencies conducted a wide range of activities and faced several challenges in addressing aquatic invasive species. Most member agencies reported conducting activities across the seven general activity categories developed by the National Invasive Species Council, including taking actions to prevent introductions of new aquatic invasive species and control the spread of existing ones (see app. III). Task Force member agencies also identified several challenges in addressing aquatic invasive species. Some of these challenges are overarching, and others relate to how member agencies plan or conduct aquatic invasive species activities specific to the activity categories.

Regarding overarching challenges, several Task Force member agencies—including officials from the Departments of the Interior and Agriculture, the Corps, and NOAA—expressed concern that their activities, though numerous, may not be adequate relative to the growing magnitude and impacts of aquatic invasive species amid decreasing or constrained agency resources. Task Force representatives further said that many of the member agencies have faced competing priorities in carrying out aquatic invasive species-related activities, with some member agencies having limited flexibility to conduct work in multiple areas. According to officials from the U.S. Geological Survey (USGS), for example, much of the agency's aquatic invasive species activities have been focused on identifying methods to treat and control Asian Carp in accordance with the Great Lakes Restoration Initiative and other funding for this work. USGS officials said that though their work on Asian Carp has been critical, it has sometimes meant that they have not been able to prioritize other needs, such as identifying marine invaders from nonballast water sources or new marine and arctic threats given the warming of ocean waters.

The following are examples of activities Task Force member agencies conducted to address aquatic invasive species along with challenges they identified related to specific activity categories, based on the responses we received to our questionnaire and interviews with officials from the Task Force and its member agencies. These examples include activities from each of the seven activity categories—(1) prevention, (2) early detection and rapid response, (3) control and management, (4) restoration, (5) research, (6) education and public awareness, and (7) leadership and international cooperation. These examples do not represent all activities conducted or challenges identified by member agencies, but rather they illustrate the nature and type of activities and challenges discussed.

Prevention

Eleven of the 13 Task Force member agencies reported conducting a range of prevention activities, often related to managing specific pathways to help prevent the introduction of aquatic invasive species into new aquatic habitats. Task Force member agencies repeatedly highlighted the importance of conducting prevention-oriented activities as a cost-effective means of addressing aquatic invasive species. Officials from some member agencies also said that they would like to conduct more prevention-oriented activities, but that they have faced challenges in doing so, in part because of policy or funding decisions within their respective agencies.[26] For example, Corps officials said they believed that it would be most cost-effective to treat certain aquatic invasive plants upstream from project boundaries before the species spreads downstream and potentially threatens project infrastructure; however, it is generally the agency's policy to treat areas within rather than outside project boundaries.[27] Some Task Force member agencies also told us that prevention activities cannot be conducted at the expense of activities aimed at controlling aquatic invasive species already established, and that a more balanced approach between prevention and control activities may be warranted.

**Prevention Efforts to Control the Spread of Quagga
and Zebra Mussels**

Several Task Force member agencies are involved in activities to prevent the spread of invasive Quagga and Zebra Mussels throughout the western United States. For example, U.S. Fish and Wildlife Service (FWS) and the National Park Service support implementation of the Quagga-Zebra Mussel Action Plan, which was developed by several state and federal agencies, as well as nongovernmental organizations in the western United States. This plan serves as a road map for identifying and prioritizing specific actions needed to prevent the further spread of Quagga and Zebra

Mussels, respond to new infestations, and manage existing ones. FWS has installed signs at National Wildlife Refuges to alert boaters about the risk of these species and has funded training in 18 states on inspecting boats and other watercraft to identify and remove the mussels. The National Park Service expended approximately $2 million in fiscal year 2014 on mussel prevention and control and monitoring at nine western parks. In addition, the Bureau of Reclamation has conducted a series of public education and outreach efforts, including the dissemination of informational pamphlets at boat shows, designed to educate the public on practices they can follow to help prevent the spread of Quagga and Zebra Mussels.

Source: Bureau of Reclamation. | GAO-16-49
Quagga Mussels covering infrastructure.

Examples of prevention activities include the following:

- **Regulations.** The U.S. Coast Guard and EPA regulate the management of ballast water—a primary pathway for the introduction of new aquatic invasive species into and within the United States— and other vessel discharges into waters of the United States. In 2012, the Coast Guard updated its ballast water regulations to include a standard for the allowable concentrations of living organisms allowed in a vessel's ballast water discharged in waters of the United States. In 2013, EPA issued a general permit that contains numeric technology-based limitations on acceptable concentrations of living organisms in ballast water discharge.[28]

- **Inspections.** FWS's Office of Law Enforcement inspects certain wildlife shipments to help ensure that prohibited species, including certain aquatic invasive species, do not enter the country. FWS's has about 120 inspectors at 49 ports of entry nationwide that review import documentation and conduct visual inspections of some shipments to help prevent species listed as injurious wildlife under the Lacey Act from being illegally brought into the country or across state lines.[29]

- **Physical barriers.** The Corps operates a series of electric barriers in the Chicago Area Waterway System located approximately 25 miles from Lake Michigan to prevent the entry of Asian Carp and other aquatic invasive species from the Mississippi River Basin into the Great Lakes. These barriers send out pulses to form an electric field in the water that discourages fish from crossing.

Early Detection and Rapid Response

Ten of the 13 Task Force member agencies reported conducting early detection and rapid response activities—activities to detect the presence of aquatic invasive species in an area and remove any newly detected species while they are localized and before they become established and spread to new areas. Aside from preventing introductions, the most cost-effective way to address an invasive species is to detect and respond to invasions early, according to documents from the U.S. Forest Service and NOAA. However, coordinated rapid response efforts have been challenging to implement due, in part, to constraints in existing funding, according to officials from some agencies. Consequently, 11 Task Force member agencies are part of a federal work group, co-led by the Department of the Interior and the National Invasive Species Council, that in January 2015 started developing a framework for a national early detection and rapid response program and a plan for an emergency rapid response fund.[30] The work group reported in July 2015 that it plans to issue a report of recommendations to implement an early detection and rapid response framework, including mechanisms for funding, to the White House and the Council on Climate Preparedness and Resilience in the fall of 2015.

Early Detection Technique Using Environmental DNA

Detection methods such as the use of environmental DNA have become widespread among Task Force member agencies, such as the U.S. Geological Survey (USGS), the U.S. Army Corps of Engineers (Corps), the

National Park Service, the Bureau of Reclamation, and the U.S Fish and Wildlife Service (FWS). Environmental DNA—genetic material shed into the environment by organisms that can be detected in samples of air, water, or soil—is a relatively new tool being used to detect invasive species, particularly in areas where the species is not abundant or is difficult to detect. For example, because they are well camouflaged in the environment, visual detection of Burmese Pythons in South Florida is difficult, with detection rates of less than 1%. Use of environmental DNA methods, however, can increase python detection rates to more than 90%, according to USGS officials. Since spring 2015, USGS researchers have been working with FWS to test water from the Loxahatchee National Wildlife Refuge in Florida to determine whether Burmese Pythons may have spread to the refuge. Although environmental DNA helps confirm the presence of an aquatic invasive species in an area, it neither confirms whether the species has become established in the area, nor does it provide information on the number or current location of any species detected.

Examples of early detection and rapid response activities include the following:

- **National early detection database.** The USGS maintains the Nonindigenous Aquatic Species Database, a publicly accessible database, to track information on the locations of aquatic invasive animals throughout the United States. Federal agencies, as well as state and local agencies and the public, can report aquatic invasive species sightings and when verified, the sightings are added to the database and updated daily by the USGS.[31]

- **Rapid response strike teams.** The FWS has five regional strike teams in place to help eradicate any new invasions as soon as possible after they are detected in the nation's 563 wildlife refuges. These strike teams survey a small portion of the acreage within national wildlife refuges when new invasions are suspected, according to FWS officials, to determine the presence of any invasions and then take actions to eradicate or contain confirmed invasions before populations spread.

Control and Management

Eleven of the 13 Task Force member agencies reported conducting activities designed to lessen and mitigate the impact or spread of aquatic invasive species on the facilities or areas they manage. Such activities may be designed to eradicate an invading species, but where eradication is not deemed feasible, such activities are designed to manage the invader by controlling the impact of the species and its spread. Activities aimed at controlling or managing the impact and spread of invasions represent a substantial portion of overall aquatic invasive species-related activities conducted, in terms of both effort and funding, according to Task Force representatives and officials from several member agencies. Some of these officials stressed the importance of sustaining efforts to control and manage aquatic invasive species to avoid reintroductions or spread of the species. For example, Corps officials said that, after eliminating infestations of Melaleuca, an invasive wetland tree, over a prescribed 10- year treatment period, periodic treatments would still be necessary to ensure new populations do not become established. Officials from several member agencies including the Corps noted, however, that limited or inconsistent funding has, at times, made it challenging to consistently manage areas as prescribed—potentially leading to the reemergence of aquatic invasive species.

Multipronged Method to Control and Manage Melaleuca

Melaleuca, an Australian tree that has destroyed many southern Florida wetlands, can be managed through a combination of biological, chemical, and physical and mechanical controls. For instance, through the introduction of weevils, a type of beetle that serves as a biological control, Melaleuca can be controlled. Researchers from the U.S. Department of Agriculture said, however, that the ability of Melaleuca trees to grow in various water depths has prevented the weevils— which require ground to burrow in—from successfully reproducing and eating the Melaleuca in swampy areas. According to National Park Service officials, Melaleuca can also be controlled if it is consistently treated over a 10-year period using the method in which the trees are first cut or hacked down with a machete or mechanical device and then sprayed with herbicides designed to kill them on the first, second, fourth, seventh, and tenth years of treatment. If this process is not followed as prescribed, however, the trees may regrow and spread. The National Park Service Exotic Plant Management Team and

Everglades National Park have contributed to control of Melaleuca in South Florida, as shown in the photo below.

Source: National Park Service. | GAO-16-49
Partially treated Melaleuca forest.

Examples of control and management activities include the following:

- **Biological controls.** To control and manage the spread of Alligatorweed, a leafy aquatic invasive plant found in the southeastern United States and California, officials from the Corps told us they are using a beetle that feeds and reproduces only on Alligatorweed. According to officials from the Corps and the U.S. Department of Agriculture, the beetle has been successful in controlling the weed, and the need for additional treatments, such as herbicide applications, has been nearly eliminated in Florida.

- **Chemical controls.** The Department of State, through the Great Lakes Fishery Commission, along with the Corps, FWS, USGS and other federal and state partners, are primarily using chemicals called lampricides to kill Sea Lamprey, an invasive fish, in their larval stage before they can attach and prey upon native fish. According to Department of State officials, as of 2015, chemical controls have led to a 90 percent reduction in the Sea Lamprey population over its historical high level.

- **Physical and mechanical controls.** The Bureau of Reclamation uses physical and mechanical control methods to remove Water Hyacinth, an aquatic invasive plant, from one of its California facilities. Bureau of Reclamation officials said that, if left untouched, Water Hyacinth clogs canals, pumps, and fish screens, which can kill the fish they are working to protect. Bureau of Reclamation officials told us that, between 2013 and 2015, they removed between 10,000 and 20,000 truckloads of Water Hyacinth from the area surrounding the facility—with a dump truck filled with Water Hyacinth leaving the facility every 5 minutes during the height of its growing season.

Restoration

Ten of the 13 Task Force member agencies reported conducting a variety of activities to restore aquatic habitats adversely affected by aquatic invasive species. Officials from a few Task Force member agencies said that it may be possible to begin restoring habitats or ecosystems while control and management activities are under way, but in some cases aquatic invasive species may need to first be controlled or contained. According to a few member agencies, this creates a challenge in that restoration activities must wait until control activities are finished, meaning that restoration may be delayed.

Examples of restoration activities include the following:

- **Habitat restoration.** NOAA reported providing funding and technical expertise for community-based habitat restoration projects, such as providing about $925,000 in 2012 for the Lower Black River Habitat Restoration Project in Ohio. The goal of this project is to restore fish and wildlife habitat in the lower Black River through actions such as the removal of aquatic invasive plants by chemical and manual techniques followed by the planting of native shrubs.
- **Native fish restoration.** The National Park Service reported removing nonnative fish from waters in a number of parks to restore native species and enhance natural aquatic biodiversity. Officials told us that they have been expending about $1 million per year since 2013 at Yellowstone National Park on lake trout removal efforts in Yellowstone Lake. These efforts include contracting with commercial fishing crews to remove invasive lake trout that have caused a significant decline in populations of the native Yellowstone Cutthroat Trout.

Research

All 13 Task Force member agencies reported conducting or sponsoring research designed to support activities to help prevent, detect, or control the impacts or spread of aquatic invasive species, as well as determine their impacts on aquatic habitats. Research is critical to identify effective techniques for prevention, detection, control, and management of aquatic invasive species and to help clarify and quantify the effects aquatic invasive species have on native species and habitats, as well as economic costs and impacts to human health, according to Task Force documents. Officials from several member agencies and Task Force representatives noted that significant gaps in knowledge in certain areas related to aquatic invasive species is a challenge and, therefore, would like to see additional research, such as a comprehensive study to identify and assess the environmental impacts and economic costs associated with invasive species in the United States. Such information is critical to understanding the magnitude of the impacts from aquatic invasive species and for obtaining funding to address problems they are causing, according to these officials. In addition, limits in scientific knowledge about newly introduced species and the levels at which they may become established or harmful, especially in ballast water, affect member agencies' ability to manage the ballast water pathway, according to officials from NOAA and the Smithsonian Environmental Research Center. Officials from the U.S. Coast Guard said that it is difficult to set regulations or establish allowable concentrations of organisms that can be safely released in ballast water when the threshold for establishment of a new potentially invasive species may not be well understood.

Federal Research on Hydrilla

Federal research on Hydrilla, a submerged invasive plant that has clogged navigation channels and other water systems across the United States, involves efforts by several Task Force member agencies. For example, the U.S. Army Corps of Engineers (Corps) conducted research on the biology of Hydrilla during 2015 to provide a better understanding of the invasion ecology of this species in northern rivers and glacial lakes. The Corps has also researched chemical treatments and application strategies to control or alter the reproduction of Hydrilla. Chemical treatments developed through research have been successful in controlling some strains of Hydrilla, according to Corps officials. Aquatic herbicides developed through research have also been successful in controlling Hydrilla, but some strains have become resistant. In addition, the Animal and Plant Health Inspection Service, in collaboration with the Corps, is researching biological controls for Hydrilla, such as releasing insects that will eat the plant.

Examples of research activities include the following:

- **Species research.** The Corps is researching various types of invasive aquatic vegetation and options for managing such species through its Aquatic Plant Control Research Program, which is authorized by statute.[32] In 2014, Corps' researchers completed field studies in Montana that used selective management strategies to control Eurasian Watermilfoil, a plant that is invasive throughout most states, including Alaska.
- **Impacts research.** Officials from USGS and NOAA have conducted research aimed at improving scientific knowledge about how aquatic invasive species may be adversely affecting ecosystems. In 2015, USGS continued research to identify whether newly established nonnative species may warrant being considered "high priority invaders," such as the Burmese Python in the Everglades. Since 2009, NOAA has conducted research to determine how certain aquatic invasive species have affected endangered salmon feeding behavior and habitat in the Pacific Northwest as part of its effort to understand the impacts that aquatic invasive species have on these native species and the ecosystems upon which they depend.
- **Pathways research.** The Maritime Administration sponsors the operation of three research facilities—in California, Maryland, and Wisconsin—that are testing the capability of treatment systems for ballast water to determine whether those systems may be approved by the U.S. Coast Guard pursuant to its ballast water regulations.

Education and Public Awareness

Eleven of the 13 Task Force member agencies reported engaging in education and public awareness activities to increase awareness about aquatic invasive species and their impacts and help minimize or prevent further introductions. According to Task Force documents, the lack of public awareness about the impacts and threats posed by some invasive species and how they are introduced is a substantial challenge for Task Force member agencies in addressing aquatic invasive species.

Lionfish Education and Public Awareness

Several Task Force member agencies are involved in raising awareness about Lionfish, a highly invasive fish that has spread throughout coastal waters of the southeast and the Caribbean. To help raise awareness, the National Oceanic and Atmospheric Administration, along with nonprofit partners, has sponsored numerous Lionfish derbies since 2010, including 10 public tournaments in 2014 in which divers could hunt the edible fish with spears. The National Park Service produced a Lionfish Response Plan in 2012 that aims to help inform the public about the Lionfish invasion and prevent and mitigate impacts to parks. Biscayne National Park, in Florida, conducts an education program in which Lionfish removed from the park are sent to classrooms for safe dissection by students. National Park Service officials told us that concentrated education efforts like this have been effective in educating the public about Lionfish. In addition, the Department of State provided funding to work with partners in the Gulf of Mexico and the Caribbean to launch a web portal that provides managers and the public with access to the latest information on Lionfish and impacts in the Atlantic Ocean.

Examples of education and public awareness activities include the following:

- **National awareness campaigns.** The Task Force, Bureau of Land Management, FWS, U.S. Forest Service, and the U.S. Coast Guard are among the federal agencies that collaborate on the "Stop Aquatic Hitchhikers!" campaign. Since 2002, this multimedia campaign has used television, billboards, and social and print media to encourage users of outdoor recreational areas to help stop the transport and spread of aquatic invasive species by, for example, making sure they clean, drain, and dry their boats and boat trailers before transporting them to different aquatic areas.

- **Local awareness events.** The National Park Service, along with state agencies and nongovernmental organizations, hosted the inaugural 5K "Race Against Invasives" run through Everglades National Park in

February 2015 to raise awareness about invasive species, especially those in Florida.

Leadership and International Cooperation

Ten of the 13 Task Force member agencies have been involved in activities to provide leadership to the aquatic invasive species community—which includes federal and nonfederal as well as international agencies working on aquatic invasive species issues—and to enhance cooperation and collaboration, such as by participating and serving as members in a range of international, national, regional, state, and local task forces, councils, and other entities. Given the often complex and widespread nature of aquatic invasive species, working across jurisdictional boundaries is the most effective approach to combating aquatic invasive species, according to Task Force officials and documents. Moreover, working with other federal and nonfederal agencies and organizations helps the Task Force to identify areas where legislation may be needed to fill gaps in statutory authority, suggest priority policy issues, and define roles and responsibilities for managing aquatic invasive species, according to Task Force documents. Officials from the regional panels told us, however, that one challenge in such work is that constrained agency funding has meant that they have not been able to consistently attend Task Force, regional panel, or other cooperative meetings.[33]

Examples of leadership and international cooperation activities include the following:

- **Aquatic Nuisance Species Task Force activities.** The Task Force conducts semiannual meetings that provide an open and public forum for members to exchange information and coordinate their aquatic invasive species activities. For example, the Task Force's May 2015 meeting included presentations on a wide range of topics, from the adoption of species-specific national management plans to recommendations from its regional panels on issues of local significance.

- **International cooperation.** Officials from the Corps and the U.S. Department of Agriculture have collaborated with scientists in China, South Korea, and Switzerland to identify and develop insect biological control agents to target invasive aquatic plants such as

Hydrilla and Eurasian Watermilfoil. For example, in fiscal year 2014, Corps officials reported expending about $450,000 on developing such control agents, which included collecting 350 plant samples from more than 90 field sites to help match invasive plants located in the United States with their countries of origin to improve the success of identifying insects to control these species.

THE TASK FORCE HAS NOT TAKEN KEY STEPS TO MEASURE PROGRESS IN ACHIEVING ITS STRATEGIC GOALS

The Task Force has not taken key steps to measure progress in achieving the goals laid out in its 2013-2017 strategic plan.[34] In 2012, the Task Force developed its 2013-2017 strategic plan, which serves to guide Task Force member agencies in conducting aquatic invasive species-related activities to implement the aquatic invasive species program. The strategic plan identifies eight goals for the program—which generally align with the seven activity categories developed by the National Invasive Species Council—as well as a number of targeted action items for Task Force member agencies to achieve these goals (see table 4).[35]

Table 4. Aquatic Nuisance Species Task Force's 2013-2017 Strategic Goals and Examples of Action Items

Goal	Examples of action items
Coordination -Maximize the organizational effectiveness of the Aquatic Nuisance Species Task Force.	• Coordinate the development and implementation of species and pathways management plans.
	• Increase communication among members, regional panels, and committees of the Task Force to prioritize issues and activities.
Prevention -Develop strategies to identify and prevent the establishment of new aquatic invasive species and slow the spread of existing ones in the waters of the United States.	• Recommend amendments to the injurious wildlife provisions of the Lacey Acta to allow a proactive approach for preventing the establishment of new invasive species through the trade of live organisms.
	• Develop and maintain a priority list for aquatic invasive species pathways.

Goal	Examples of action items
Early detection and rapid response -Identify and respond to aquatic invasive species within a timely manner following introduction to prevent their establishment and/or spread.	• Increase public and industry involvement in reporting sightings of aquatic invasive species through the U.S. Geological Survey's online reporting form for its publicly available database on locations of aquatic invasive species. • Develop a rapid response technical support network that can provide resources and technical support in response to newly detected species.
Control and management - Control established aquatic invasive species when feasible and when the benefits of managing the established species outweigh the costs of removing them with respect to harm to the environment, the economy, and public health.	• Increase the number of training workshops for natural resource managers and the total number of personnel and volunteers trained in control measures for aquatic invasive species. • Identify gaps in control efforts and tools.
Restoration -Protect and rehabilitate native species and ecosystems by conducting habitat restoration efforts on multiple scales.	• Ensure that federal land and water management guidance manuals consider aquatic invasive species issues during the planning and development of habitat restoration projects. • Compile, highlight, and share lessons learned for both restoration successes and failures within the United States.
Education and outreach - Increase awareness concerning the threats of aquatic invasive species, emphasizing the impacts, importance of prevention and containment, and recommendations for appropriate domestic and international actions.	• Utilize the internet and social media as well as traditional media sources to disseminate information and promote awareness of aquatic invasive species. • Cooperate with media outlets to reach a broad range of the public with aquatic invasive species messages.

Since its inception, the Task Force has provided one report to Congress, in 2004.

Task Force representatives said they expect to finalize and issue their draft report by the end of 2015. In reviewing a draft of the report, we found that the draft provided an overview and examples of aquatic invasive species activities conducted by the Task Force, member agencies, regional panels, and states since the Task Force's 2004 report, as well as some information on the role of Task Force member agencies in aquatic invasive species management. After they finalize the 2015 report, Task Force representatives have not indicated that they would begin submitting reports annually to meet this reporting requirement in the future.

Task Force representatives also said they have no plans to develop an operational plan, as called for in the strategic plan, but acknowledged the importance of developing a means to regularly track various member agencies' aquatic invasive species activities and measure progress toward meeting the strategic goals. Specifically, in response to our inquiry into the status of an operational plan, Task Force representatives told us in May 2015 that they planned to discuss the possibility of reviving or modifying the reporting matrix they had used in 2012. Task Force representatives subsequently told us that, during a June 2015 meeting, member agencies agreed that a tracking mechanism was important. However, they also told us that they did not determine what such a mechanism would look like, how it would be implemented and by whom, or how to address concerns expressed by some member agencies that the mechanism not burden agency staff already working at capacity in light of constrained funding. Task Force representatives said they plan to further discuss the idea of reviving or modifying the reporting matrix at their next semiannual Task Force meeting in November 2015. But, representatives could not tell us when they planned to make a decision on the approach they would take or provide specifics on what information they would collect or how they would measure progress in achieving their strategic goals.

By developing and regularly using a tracking mechanism—that would include the elements envisioned for an operational plan and required by the 1990 Act—the Task Force could better position itself to (1) measure progress in achieving its strategic goals and (2) comply with certain requirements in the 1990 Act for the aquatic invasive species program. Addressing aquatic invasive species is a complex, interdisciplinary issue with the potential to affect many sectors and levels of government operations. Strategic planning is a way to respond to this governmentwide problem on a governmentwide scale. Our past work on crosscutting issues has found that governmentwide strategic planning can integrate activities that span a wide array of federal, state, and local entities, as well as provide a comprehensive framework for making resource decisions and holding agencies accountable for achieving strategic goals.[41] With its strategic plan, the Task Force has a framework in place to guide and integrate the numerous and varied aquatic invasive species activities spanning many member agencies. In addition to measuring progress in achieving the Task Force's strategic goals, developing and regularly using a tracking mechanism could also help the Task Force meet the 1990 Act's requirements to describe its members' roles and specific activities and to report annually to Congress on the program's progress.

CONCLUSION

Aquatic invasive species, a serious and growing problem affecting all states and U.S. territories, have been likened to a never-ending oil spill, given that they are notoriously difficult to eradicate once they become established. Though hard to calculate, the economic and ecological harm caused by aquatic invasive species is vast. Capturing how much federal agencies have expended—and will likely need to expend—to effectively address aquatic invasive species is also challenging. Consequently, it is not possible to identify how much may be needed to fully address aquatic invasive species, both in terms of current invasions or measures to prevent future invasions. Capturing how much progress federal agencies have made in combatting aquatic invasive species is similarly challenging.

The Task Force and its member agencies have taken significant steps—including conducting a wide array of activities and developing a strategic plan to guide their efforts—to address the threats and impacts of aquatic invasive species. However, the Task Force has not met several of the 1990 Act's requirements, including reporting annually to Congress on the program's progress, or developed a mechanism to ensure its strategic goals are measurable and accountable, such as through an operational plan, as called for in its strategic plan, because of constrained funding and limited resources. Task Force member agencies agreed that a mechanism to track activities and measure progress was important, but the Task Force has not decided what the mechanism would look like, how it would be implemented and by whom, or how to address concerns that it not burden agency staff already working at capacity. Developing and regularly using a tracking mechanism could help the Task Force measure progress in achieving its strategic goals, as well as help the Task Force meet the 1990 Act's requirements to describe its members' roles and specific activities and to report annually to Congress on the program's progress. Moreover, such a mechanism could provide a starting point for identifying funding gaps and priorities, better positioning the Task Force to meet the 1990 Act's requirement to include recommendations for funding to implement elements of its aquatic invasive species program.

RECOMMENDATION FOR EXECUTIVE ACTION

As the Aquatic Nuisance Species Task Force considers how to measure progress toward accomplishing its strategic goals, we recommend that the Task Force develop and regularly use a tracking mechanism, to include elements envisioned for an operational plan and to largely meet requirements in the 1990 Act, including:

- specifying the roles of member agencies related to its strategic plan,
- tracking activities to be conducted by collecting information on those activities and associated funding,
- measuring progress member agencies have made in achieving its strategic goals, and
- reporting to Congress annually on the progress of its program.

AGENCY COMMENTS AND OUR EVALUATION

We provided the Secretaries of Agriculture, Commerce, Defense, Homeland Security, Interior, State, and Transportation and the Administrator of the EPA a draft of this report for their review and comment. Only the Department of the Interior and the Department of Commerce's NOAA provided written comments. Interior generally agreed with the report's findings and recommendation, and NOAA disagreed, as further discussed below. The Department of Defense's U.S. Army Corps of Engineers, the Department of State, and EPA indicated that they had no comments on our report through e-mail communications provided through departmental audit liaisons on October 19, October 21, and October 23, 2015, respectively. We also received e-mails provided through audit liaisons from the following departments that stated that the departments agreed with the report's findings and recommendation and had no other comments: The Department of Agriculture's Animal and Plant Health Inspection Service and U.S. Forest Service (dated October 29, and October 30, 2015, respectively); the Department of Transportation (dated October 26, 2015); and the Department of Homeland Security (dated October 15, 2015).

In its written comments, the Department of the Interior stated that it generally agreed with the findings of our report and concurred with our recommendation. Interior stated that it appreciated our review of the challenges faced by the Task Force in addressing and managing risks posed by the introduction and proliferation of aquatic invasive species. Interior stated that the Task Force, of which its FWS is a co-chair, is currently evaluating the reporting matrix to improve its utility as a tracking mechanism. Additionally, Interior stated that, at its November 2015 meeting, the Task Force agreed to track accomplishments using a modified activity tracking tool while its members continue to evaluate how best to track their activities going forward. Interior also stated that the Task Force's report to Congress is undergoing final agency review, and it is expected to be delivered to Congress in the coming months, which, together with its tracking efforts, will help provide the Task Force with a mechanism to both measure and communicate progress toward its strategic goals, as called for in our report. We agree that using a modified activity tracking tool and completing the report to Congress will be positive first steps in the Task Force's measuring progress toward accomplishing its strategic goals and meeting requirements in the 1990 Act, in accordance with our recommendation. Interior also provided technical comments, which we incorporated, as appropriate.

In its written comments, NOAA disagreed with several aspects of our findings, conclusions, and recommendation. In addition, NOAA stated that our report did not sufficiently address certain aspects of the mandate to conduct the review contained in section 1039(a)(2) of the Water Resources Reform and Development Act of 2014. First, NOAA stated that the report did not mention future costs to mitigate the impacts of aquatic invasive species and that, although it may be difficult to give specific numbers, some information could be speculated upon. In the opening paragraph of our report, we state that the impacts of invasive species in the United States are widespread and expected to increase, with profound consequences for the economy and the environment. We cite a 2005 academic study—the most recent comprehensive study of its kind— that estimates the environmental impacts and economic costs associated with invasive species at almost $120 billion per year. Additionally, through our questionnaire, we requested that federal member agencies provide planned activities and estimated expenditures for future years. However, as we describe in the scope and methodology appendix (app. I) of our report, we decided not to report future estimated expenditures given the limited information provided by some member agencies. We believe that reporting partial information could be misleading and could underestimate likely future expenditures.

Second, NOAA stated that our analysis could have gone into more detail about current federal spending on prevention activities. We limited our reporting of expenditures for fiscal years 2012 through 2014 to estimates of *total* annual expenditures for each Task Force member agency because many member agencies reported that they could not provide estimates of their expenditures by activity category, including prevention. Third, NOAA stated that we did not address whether federal spending is adequate for the maintenance and protection of services provided by federal facilities. As we note in our report, capturing how much federal agencies have expended—and will likely need to expend—to effectively address aquatic invasive species is challenging. Given the limited information available from the Task Force member agencies on current and planned expenditures related to aquatic invasive species, we determined we would not be able to reliably conduct an analysis of the adequacy of federal spending. Lastly, NOAA stated that we chose to focus on the Aquatic Nuisance Species Task Force and its strategic plan rather than documenting other legislative and programmatic efforts that target the prevention, control, and management of aquatic invasive species. The scope of our review includes all federal member agencies of the Task Force, and in discussing activities and challenges those member agencies face in addressing aquatic invasive species, our report highlights many of the legislative and programmatic efforts those agencies are undertaking, such as efforts by the U.S. Coast Guard and EPA to regulate and manage ballast water through updated regulations.

NOAA also stated that our report did not mention federal mandates intended to address aquatic invasive species other than the Nonindigenous Aquatic Nuisance Prevention and Control Act of 1990, the National Invasive Species Act of 1996, and Executive Order 13112. NOAA stated that at its exit conference with us on July 14, 2015, it noted that many federal agencies receive additional directions or mandates to address or respond to aquatic invasive species and their impacts and that each agency must balance these mandates. We agree that federal agencies may have multiple responsibilities in addressing aquatic invasive species—we outline many of these responsibilities in table 1 of the background of our report where we describe the key roles and responsibilities of Task Force member agencies under various federal laws. Also, in describing examples of the activities and challenges member agencies face in addressing aquatic invasive species in the second objective of our report, we identify and describe many of the requirements and mandates member agencies must follow. For example, we describe efforts of the FWS' Office of Law Enforcement to enforce the Lacey Act, which prohibits the importation and interstate transport of wildlife listed as injurious, among other things. NOAA also stated that balancing and responding to various requirements ultimately affects the agencies' ability to adequately respond to this national issue. We agree with this statement, and in our discussion of challenges faced by member agencies in addressing aquatic invasive species, we report that many of the member agencies have faced competing priorities in carrying out aquatic invasive species-related activities, with some member agencies having limited flexibility to conduct work in multiple areas.

In addition, NOAA stated that the reported presence of a species in USGS' Nonindigenous Aquatic Species database (one of two key sources we used to prepare species' location information for the map) does not mean that the species is established in a particular state's waters as the map portrays. In our draft report, in a note to the figure, we included a statement to clarify that species distributions in the map represent the reported presence of a species in at least one, but not necessarily all, bodies of water in the state, and do not necessarily indicate establishment of the species in any part of the state. To further clarify this point so as not to potentially mislead readers, in response to NOAA's comment, we have updated the figure title and note and also added a statement to this effect in the body of the report. Second, NOAA stated that Caulerpa, one aquatic invasive species we highlighted in the figure, had been eradicated. Upon receipt of this information from NOAA and in light of obtaining additional supporting data, we removed Caulerpa from the figure. Third, NOAA stated that providing points of pathways of invasion was confusing or inaccurate in some cases. We agree that the manner in which we linked our description of the pathways of invasion to the map in the draft report could be misinterpreted; consequently, in response to NOAA's comment, we disassociated the description of pathways from the map. We believe that providing a description of various pathways aquatic invasive species may use to enter and spread into new areas is important context for our report.

Furthermore, concerning our recommendation that the Task Force develop and regularly use a tracking mechanism, to include elements envisioned for an operational plan and to largely meet requirements in the 1990 Act, NOAA stated that it does not believe the recommendation can address problems faced by the Task Force. NOAA stated that, with respect to measuring progress, the Task Force agreed to use an activity matrix to compile information, but the matrix has not been updated since 2012 for several reasons, including because of uncertainties in funding, shifting priorities, and the loss of the Task Force Executive Secretary position, which has not been filled since the former Executive Secretary retired in 2013. NOAA further stated that the report does not address the underlying causes that have hindered Task Force efforts to track progress, including the limited budget under which the Task Force operates, which has been reduced significantly in recent years. Our recommendation was not intended to comprehensively address the problems faced by the Task Force, but rather was more narrowly focused. Specifically, the intent of our recommendation is to help the Task Force regularly track progress toward achieving its strategic goals in a manner that ensures it also largely meets requirements in the 1990 Act, such as reporting to Congress annually on the progress of its program. In our report, we discuss the constrained funding environment and limited resources the Task Force and its member agencies reported working under, including having limited staff devoted directly to the Task Force and facing the constrained funding environment that emerged from sequestration in 2013 and 2014. We believe that by implementing our recommendation—that is, by developing and regularly using a tracking mechanism to include the roles of member agencies, activities conducted and associated funding, and progress made in achieving strategic goals—the Task Force would be in a better position to identify and communicate its progress, as well as funding or resource needs to address problems faced by the Task Force. As we note in our report, capturing how much federal agencies have expended—and will likely need to expend—to effectively address aquatic invasive species is challenging. But by developing and regularly using a tracking mechanism, we believe the Task Force would be better-positioned to assess funding gaps and priorities and begin to identify solutions to address the challenges member agencies face in addressing aquatic invasive species.

Finally, NOAA identified examples where it stated information portrayed in our report could have evolved into recommendations. For example, NOAA commented that a recommendation that calls for a more balanced approach in conducting prevention activities would be beneficial. In our report, we state that member agencies repeatedly highlighted the importance of conducting prevention-oriented activities as a cost-effective means of addressing aquatic invasive species. We also note that officials from some member agencies said they would like to conduct more prevention-oriented activities, but that prevention activities cannot be conducted at the expense of activities aimed at controlling aquatic invasive species already established, and that a more balanced approach between prevention and control activities may be warranted. We include this and the other examples NOAA references in our report to provide context on an issue, provide examples of activities being undertaken by member agencies, or describe challenges faced by member agencies in addressing aquatic invasive species—consistent with the objectives and scope of work conducted for this review. Consistent with government auditing standards, we are to have sufficient, appropriate evidence to provide a reasonable basis for findings and conclusions before we can develop recommendations. Based on our work, we did not have sufficient evidence to provide a reasonable basis for making recommendations on the examples NOAA identified. We encourage NOAA to continue to work with Task Force member agencies and others to pursue areas they identify as needing additional work, such as identifying ways to take a more balanced approach across prevention and control activities. We believe that by implementing our recommendation, NOAA, as one of the co-chairs of the Task Force, would be in a better position to identify funding gaps and priorities, and determine recommendations for funding based on emerging needs.

NOAA also provided technical comments, which we incorporated, as appropriate.

Anne-Marie Fennell
Director,
Natural Resources and Environment

APPENDIX I: OBJECTIVES, SCOPE, AND METHODOLOGY

This report examines (1) how much the Aquatic Nuisance Species Task Force (Task Force) member agencies expended addressing aquatic invasive species from fiscal year 2012 through 2014; (2) activities conducted by Task Force member agencies and challenges in addressing aquatic invasive species; and (3) the extent to which the Task Force has measured progress in achieving the goals of its 2013-2017 strategic plan.

For all three objectives, we reviewed aquatic invasive species-related laws, including the Nonindigenous Aquatic Nuisance Prevention and Control Act of 1990, as amended (the 1990 Act),[1] regulations, and academic studies.

We conducted interviews with, and obtained documentation from, the co-chairs of the Task Force and other Task Force representatives; officials from the 13 Task Force federal member departments and agencies (member agencies);[2] and representatives from each of the Task Force's six regional panels to learn about their roles and responsibilities, aquatic invasive species-related activities, and any expenditure information they maintain related to those activities. In addition, we interviewed staff from the National Invasive Species Council to learn about their efforts to collect information on federal expenditures for invasive species activities.

To determine how much Task Force member agencies expended addressing aquatic invasive species for fiscal years 2012 through 2014 and obtain information on activities conducted, we developed and disseminated a questionnaire to the 13 Task Force member agencies, requesting information on their estimated expenditures and activities conducted to address aquatic invasive species. Specifically, the questionnaire requested member agencies to provide estimates of their expenditures for the activities they conducted in each of the following seven aquatic invasive species activity categories: (1) prevention, (2) early detection and rapid response, (3) control and management, (4) research, (5) restoration, (6) education and public awareness, (7) and leadership and international cooperation.[3] These were the same activity categories used by the National Invasive Species Council to collect and report information for its annual invasive species interagency "crosscut" budget summary. The council's annual budget summary includes estimates of federal agency expenditures and planned funding on activities to address all types of invasive species, but it does not include a breakdown of expenditure by type, including expenditures specific to aquatic invasive species. Therefore, the council's annual budget summary provided a framework for us to follow in developing our questionnaire, but we could not use information from the budget summary to obtain or report information on federal expenditures specific to aquatic invasive species. Several Task Force member agency officials recommended that we follow the council's framework for our questionnaire since many of the member agencies provide information to the council, and they suggested that following a similar framework would facilitate their ability to respond to our request. In developing our questionnaire, we worked with staff from the National Invasive Species Council and conducted pretests with three member agencies to obtain their comments, which were incorporated as appropriate.

In our questionnaire, we requested that each member agency provide (1) its estimated expenditures for fiscal years 2012 through 2014 (the most recent years for which member agencies reported reliable data were available), (2) examples of aquatic invasive species activities conducted during this time period, and (3) its planned activities and estimated expenditures for future years, which we defined as fiscal years 2015 and 2016. We also included questions about how the Task Force member agencies prepared their estimates, their sources of information, any challenges or limitations in preparing the estimates, and whether the estimates were reviewed by their budget or financial offices.

We received completed responses from all 13 of the Task Force member agencies. The member agencies provided information on their activities conducted to address aquatic invasive species, but member agencies varied in the level of detail they provided about their estimated expenditures. Twelve of the 13 member agencies included at least some information on their estimated expenditures for fiscal years 2012 through 2014, but the U.S. Forest Service reported that it was unable to provide estimates. For the other 12 agencies, they varied in their ability to provide consistent and complete information on their estimated expenditures at the level of detail we requested in our questionnaire. With respect to the expenditure information for fiscal years 2012 to 2014, some agencies were able to provide estimates of their expenditures by activity category, but many reported that they could not provide estimates at this level of detail. For example, the Environmental Protection Agency reported its expenditures supported activities for five of the seven activity categories, but because it could not provide separate estimates for each of these categories it reported all of its expenditures under the prevention category. Similarly, the National Park Service reported conducting activities in all seven activity categories in fiscal years 2012 and 2013, but provided estimates for two activity categories (research and restoration) and reported that it was unable to determine how much of its estimated expenditures went toward the other five activity categories in these years. Based on inconsistencies and incomplete responses across the 13 member agencies, we decided to limit our reporting for fiscal years 2012 through 2014 to estimates of total annual expenditures for each Task Force member agency.

With respect to future expenditures for fiscal years 2015 to 2016, a few member agencies indicated they did not have estimates of expenditures for future years, though others had partial estimates. To avoid reporting potentially misleading information that could underestimate likely future expenditures compared to amounts reported for fiscal years 2012 through 2014, we decided not to report the future expenditure estimates provided to us. Similarly, 9 of the 13 member agencies reported that they were not able to provide estimates for how much they expended addressing specific aquatic invasive species, citing reasons such as expenditures being tracked at a project level rather than by a specific species. Therefore, we do not include species-specific expenditure information in our report.

After receiving completed questionnaires, we followed up with Task Force member agency officials to obtain clarification or additional information, as needed. We did not independently verify the accuracy of the estimated expenditures reported by the member agencies, which likely include some over- and some under-estimates. For example, in its response, the U.S. Fish and Wildlife Service (FWS) described various activities that were implemented through projects supported with grant funding from the Wildlife Sport Fish Restoration Program.[4] But, FWS did not include expenditure estimates for these project activities because it could not reliably estimate how much of the grant funding should be attributed to the aquatic invasive species component of the grant-funded projects. We asked each of the Task Force member agencies for their assessment of whether their estimated expenditures for fiscal years 2012 to 2014 were an underestimate, overestimate, or about right. Ten of the member agencies responded that their estimates were "about right," and two indicated they were underestimates (one member agency did not provide estimates). Accordingly, the expenditures reflect the agencies' best estimates of how much they expended on aquatic invasive species activities during these years. Based on our assessment of these responses, along with the responses provided through the questionnaire, we determined that the expenditure estimates for fiscal years 2012 through 2014 were sufficiently reliable for purposes of this report—to provide general estimates of total annual expenditures by Task Force member agencies on activities to address aquatic invasive species.

To describe the activities conducted by Task Force member agencies and any challenges in addressing aquatic invasive species, we built on the information gathered through our questionnaire and conducted a series of interviews with officials from the 13 member agencies, the federal ex-officio member of the Task Force (the Smithsonian Environmental Research Center), and each of the Task Force's six regional panels. Through these interviews, we collected information and documentation on aquatic invasive species activities conducted and any challenges agencies identified in addressing aquatic invasive species. Many of the activities and challenges relate to ongoing activities that span multiple fiscal years and thus the information we collected often highlights, but is not limited to, fiscal years 2012 through 2014. We also conducted site visits in Southern Florida, Northern California, and Western

Washington to interview local federal officials and observe activities at the sites, such as inspections of shipments of live fish to search for aquatic invasive species and research being conducted at research facilities. We selected these locations based on the number and variety of aquatic invasive species and federal agencies, as well as the types of activities conducted in those locations. Information we obtained from our interviews and site visits on activities conducted and challenges identified are not generalizable, but we believe the examples we obtained provide important insights into the wide array of aquatic invasive species activities being undertaken across the 13 Task Force member agencies and the challenges agencies face in conducting those activities.

To determine the extent to which the Task Force has measured progress in achieving the goals of its 2013-2017 strategic plan, we conducted interviews with and obtained documentation from Task Force representatives, officials from the 13 Task Force member agencies, and officials representing the six regional panels. We reviewed the Task Force's 2013-2017 strategic plan, its 2012 reporting matrix, and other documentation related to the Task Force's efforts to collect information related to its strategic plan. We then analyzed and compared this information to program requirements identified in the 1990 Act,[5] our previous reports on leading practices provided by the GPRA Modernization Act of 2010,[6] and our executive guide on strategic planning,[7] as appropriate.

We conducted this performance audit from November 2014 to November 2015 in accordance with generally accepted government auditing standards. Those standards require that we plan and perform the audit to obtain sufficient, appropriate evidence to provide a reasonable basis for our findings and conclusions based on our audit objectives. We believe that the evidence obtained provides a reasonable basis for our findings and conclusions based on our audit objectives.

APPENDIX II: EXAMPLES OF AQUATIC INVASIVE SPECIES AND THEIR REPORTED PRESENCE BY STATE, AND COMMON PATHWAYS

Table 5 provides descriptions of the aquatic invasive species used as examples, and table 6 provides descriptions of common pathways of invasion.

Table 5. Description of Examples of Aquatic Invasive Species

Species	States affected	Species description
Common Water Hyacinth	Alabama; Arizona; Arkansas; California; Colorado; Connecticut; Delaware; Florida; Georgia; Hawaii; Illinois; Kentucky; Louisiana; Massachusetts; Mississippi; Missouri; New Jersey; New York; North Carolina; Oregon; South Carolina; Tennessee; Texas; Virginia; Washington	A free-floating flowering plant that forms dense colonies that block sunlight and crowd out native species, and can clog water intake structures, canals, and irrigation systems, therefore damaging ecosystems, raising operation costs, and impacting navigation of waterways. Common Water Hyacinth has become established in the Southeast, Northeast, and along the Pacific Coast of the United States after being introduced from South America through the ornamental aquarium trade.
Chinese Mitten Crab	California; Connecticut; Delaware; Louisiana; Maryland; Michigan; New Jersey; New York; Ohio; Oregon; Washington	Crustaceans that threaten fisheries and aquatic ecosystems, cause clogging at water intake structures, and have the potential to transport disease to humans. Chinese Mitten Crabs are native to the Pacific Coasts of China and Korea. They were introduced to California, the Great Lakes, and the Mid-Atlantic Coast through ballast water discharge and the intentional or accidental release of these crabs purchased as food. Chinese Mitten Crabs are listed as an injurious species under the Lacey Act, making it illegal to import or transport them between states.
Sea Lamprey	Illinois; Indiana; Michigan; Minnesota; New York; Ohio; Pennsylvania; Wisconsin	Fish that prey on native fish by attaching to the outside of a native host fish and draining its nutrients, often killing the host fish. Sea Lamprey can devastate native fish populations and harm fishing industries. They are native to the Atlantic Ocean, but invaded the Great Lakes by passing through canals.

Species	States affected	Species description
Asian Carp	Every state but Alaska, Montana, and Rhode Island	Collectively refers to four species: Grass Carp, Bighead Carp, Black Carp, and Silver Carp. These fish harm ecosystems, threaten native species, and are also a danger to human health when they jump out of the water, which can injure or distract recreational boaters. Found naturally in Russia and China, each species of Asian Carp was intentionally introduced in the United States to improve water systems. Specifically, Grass Carp were intended for vegetation control, Bighead and Silver Carp for water quality improvement, and Black Carp for aquaculture-related work on parasite control. However, these fish escaped their introduction areas and, as a group, Asian Carp have invaded lakes in nearly all U. S. states, including the invasion of Grass Carp into the Great Lakes. Bighead Carp, Black Carp, and Silver Carp are listed as injurious species under the Lacey Act, making it illegal to import or transport them between states.
Zebra and Quagga Mussels	Alabama; Arizona; Arkansas; California; Colorado; Connecticut; Delaware; Georgia; Illinois; Indiana; Iowa; Kansas; Kentucky; Louisiana; Maryland; Massachusetts; Michigan; Minnesota; Mississippi; Missouri; Nebraska; Nevada; New Jersey; New York; North Dakota; Ohio; Oklahoma; Pennsylvania;	Distinct species of mussels that cause similar types of damage to native ecosystems by destroying native fish habitat and food webs. Both species of mussels colonize on water supply pipes, which can impact hydroelectric and nuclear power plants, public water supply plants, and industrial facilities. These mussels also affect navigation and boating by increasing the weight on vessels and sinking navigation buoys.

Table 5. (Continued)

Species	States affected	Species description
	South Dakota; Tennessee; Texas; Utah; Vermont; Virginia; West Virginia; Wisconsin	
		They arrived in the United States from the Black, Caspian, and Azov Seas through ballast water discharge and have spread across much of the continental United States from recreational boats and fishing gear. Zebra Mussels (but not Quagga Mussels) are listed as an injurious species under the Lacey Act, making it illegal to import or transport them between states.
Northern Snakehead	Arkansas; California; Delaware; Florida; Illinois; Maryland; Massachusetts; New Jersey; New York; North Carolina; Pennsylvania; Virginia	A fish that competes with native species for food and habitat and can eat small birds and mammals in addition to other fish and aquatic reptiles. It originates from China, Russia, and Korea and has invaded numerous areas of the north and Mid-Atlantic Coast of the United States as well as Florida, Illinois, Arkansas, and California. The U.S. Fish and Wildlife Service estimates that snakeheads arrived in U.S. waters as an import for the live food market and became invasive by intentional release from the aquarium trade. Northern Snakehead is listed as an injurious species under the Lacey Act, making it illegal to import or transport them between states.

Species	States affected	Species description
Nutria	Alabama; Arizona; Arkansas; California; Delaware; Florida; Georgia; Idaho; Iowa; Kansas; Kentucky; Louisiana; Maryland; Minnesota; Mississippi; Missouri; Nebraska; Nevada; New Mexico; New York; North Carolina; Ohio; Oklahoma; Oregon; Pennsylvania;	An aquatic invasive mammal, resembling a beaver, which has destroyed habitat and threatened endangered species in many states. Nutria overgraze wetland habitats, destroying ecosystems. It is native to southern South America and was originally imported into Louisiana for fur farming, but escaped to areas including Louisiana, Maryland, Oregon, Texas, and Washington.
	Tennessee; Texas; Virginia; Washington; West Virginia	In Maryland, populations of Nutria have largely been eliminated, according to Animal and Plant Health Inspection Service officials.
Snowflake Coral	Hawaii	A soft, nonreef forming coral that grows over hard surfaces, but also grows quickly over native corals and reef surfaces, which kills native reefs, especially the native black coral that supports a vibrant economy and ecosystem in Hawaii. Snowflake Coral also feeds on zooplankton that supports reef ecosystems. It is native to the Caribbean and was introduced through hull fouling to Hawaii.
Water Lettuce	Arizona; California; Colorado; Connecticut; Delaware; Florida; Georgia; Hawaii; Kansas; Louisiana; Maryland; Mississippi; Missouri; New Jersey; New York; North Carolina; Ohio; South Carolina; Texas	A floating aquatic invasive plant that grows into dense mats that can clog waterways and water intake structures. Scientists disagree as to whether the plant is native to the continental United States or originated from South America or Africa, but agree that it was likely transported by ballast water across the United States and is invasive in Hawaii.

Table 5. (Continued)

Species	States affected	Species description
New Zealand Mudsnail	Arizona; California; Colorado; Idaho; Illinois; Minnesota; Montana; New York; Nevada; Ohio; Oregon; Pennsylvania; Utah; Washington; Wisconsin; Wyoming	A small freshwater snail that can cause declines in native snail populations by out competing them and can alter stream ecosystems through their rapid spread. The New Zealand Mudsnail originates from New Zealand and adjacent small islands and has become established in the western United States and the Great Lakes. It is believed that the snail was introduced to the Great Lakes and western states through ballast water discharge and hull fouling.
Alligatorweed	Alabama; Arkansas; California; Florida; Georgia; Illinois; Kentucky; Louisiana; Mississippi; North Carolina; Oklahoma; South Carolina; Tennessee; Texas; Virginia	A leafy aquatic plant that forms dense mats that crowd out native species and impedes recreational activities such as boating, swimming, and fishing. Alligatorweed originates from South America and was introduced through ballast water discharge in the southeastern United States and California.
Hydrilla	Alabama; Arizona; Arkansas; California; Connecticut; Delaware; Florida; Georgia; Indiana; Iowa; Kentucky; Louisiana; Maine; Maryland; Massachusetts; Mississippi; New Jersey; New York; North Carolina; Pennsylvania; South Carolina; Tennessee; Texas; Virginia; Washington	A submersed aquatic plant that can grow in thick mats and has the ability to regrow from small pieces of the plant. It can clog canals, irrigation, and structure vents. Hydrilla is listed as a Federal Noxious Weed, which restricts its importation, exportation, and interstate movement. According to U.S. Army Corps of Engineer officials, bacteria can grow on the surface of Hydrilla, such as Aetokthonos hydrillicola, and kill some waterfowl and Bald Eagles when they ingest the bacterium. Hydrilla is found in the Southeast, Northeast, Washington, California, Arizona, Indiana, and Iowa. It was introduced from Asia through the aquarium trade.

Species	States affected	Species description
Lionfish	Alabama; Florida; Georgia; Louisiana; Mississippi; New Jersey; New York; North Carolina; Rhode Island; South Carolina; Texas; Virginia	A saltwater fish that preys on native fish communities and damages fragile coral ecosystems. Lionfish have several spines that contain potent venom that can administer a painful sting to humans and potential predators. As a result, it has few natural predators outside its native habitat. Lionfish are native in the western Pacific Ocean, Indian Ocean, and Red Sea; they are invasive in the western Atlantic Ocean, Caribbean Sea, and Gulf of Mexico. The arrival of Lionfish in the Atlantic Ocean is attributed to the aquarium trade and intentional release of pet Lionfish.
Melaleuca	Florida; Hawaii; Louisiana	An invasive tree that crowds out native species and causes damage to ecosystems, especially delicate wetlands. Melaleuca has adapted to grow at varying levels of water and each tree can hold millions of seeds, which allow this invasive tree to spread rapidly. It is originally from Australia and was brought to the United States as an ornamental plant used for erosion control. It is found in Hawaii, Florida, and Louisiana and is listed as a Federal Noxious Weed, which restricts its importation, exportation, or interstate movement.
Burmese Python	Florida	A constrictor snake that preys on and competes with native species. It is native to Southeast Asia and is invasive in Southern Florida. It was introduced by pet owners through escape or intentional release. The Burmese Python is listed as injurious wildlife under the Lacey Act, making it illegal to import or transport the snake between states.

Sources: USGS; NOAA; USFWS; EPA; USDA; Hawaii Invasive Species Council; Nonindigenous Aquatic Species Database; PLANTS Database; Global Invasive Species Database; GAO. | GAO-16-49

Table 6. Examples of Common Pathways

Common pathways	Description of common pathways
Trade	The transport of items through commerce or other exchanges—by boat, airplane, the mail, or other transportation methods—increases the probability of introducing aquatic invasive species. Aquatic invasive species can be intentionally sent through trade or mail routes such as with Chinese Mitten Crabs for human consumption, or they can hitchhike by various transportation modes or on other species.
Ballast water	Water in a ship's holding tank that is used for stability and safety. When ballast water is acquired in one body of water or location and released in another, the potential for invasive species to be released and impact coastal communities is immense. Ballast water discharge is a common pathway through which aquatic invasive species are introduced into U.S. waters. Ballast water can introduce aquatic invasive species at any port in fresh or saltwater around the United States. For example, the introductions of Quagga and Zebra Mussels have been attributed to ballast water discharges.
Hull fouling	The accumulation of organisms on a vessel's exterior surfaces can occur on commercial, recreational, or military vessels. Also called biofouling, hull fouling can introduce aquatic invasive species into any water body where boats or other water vessels are used and species become detached from surfaces. Snowflake Coral, for example, was introduced to Hawaii through this pathway.
Outdoor gear and bait	Fishing and hiking gear, as well as live bait, from outdoor activities has been known to spread aquatic invasive species to new waters. Fishing equipment, diving gear, and other recreational items that are transported among water bodies increase the risk of introduction. Discarding unused live bait after fishing can also introduce species that disrupt their new ecosystems and eliminate competing native species. For example, New Zealand Mudsnails have been unintentionally transported on fishing equipment across water bodies throughout the United States.
Boats and boat trailers	Recreational boats and boat trailers are a common pathway of introduction for aquatic invasive species, such as Hydrilla. Small pieces of this plant can survive on trailers and boat motors and regrow in new locations. Boat trailers are a common pathway for species to move from one water body to another and numerous federal and state agencies are invested in preventing species'

	spread through this pathway through the "Stop Aquatic Hitchhikers!" campaign and cleaning stations at boat launches across the United States. Boats and boat trailers, for example, are estimated to have spread Quagga and Zebra Mussels from the Great Lakes to other water bodies in the United States.
Common pathways	Description of common pathways
Aquarium release	Escape or intentional release of unwanted pets and ornamental or aquarium plants, as well as escape or intentional release of species from aquaculture farms, can be a source of invasive species in all parts of the country. Lionfish and Burmese Pythons, for example, were introduced to U.S. waterways after being released from aquaria.
Intentional introductions	Illegal stocking and the live food industry can introduce aquatic invasive species. Although it may be prohibited by law, people release fish into new waters for sport fishing, which can have severe impacts on ecosystems. The import of live, exotic plants and animals and the release of those species in new locations to create local live food sources is another pathway of intentional introduction. The Northern Snakehead, for example, is thought to have been introduced in Maryland from the live food industry after it became a popular food source.

Sources: USGS; NOAA; USFWS; EPA; USDA; Hawaii Invasive Species Council; Nonindigenous Aquatic Species Database; PLANTS Database; Global Invasive Species Database; GAO. | GAO-16-49

APPENDIX III: AQUATIC INVASIVE SPECIES ACTIVITIES CONDUCTED BY TASK FORCE MEMBER AGENCIES

Through our questionnaire to the 13 federal member agencies of the Aquatic Nuisance Species Task Force (Task Force), we requested that member agencies identify the types of aquatic invasive species activities they conducted during fiscal years 2012 through 2014, including how those activities fell within the seven general activity categories developed by the National Invasive Species Council. The Task Force member agency responses are summarized in table 7.

Table 7. Aquatic Invasive Species Activities Conducted by Task Force Member Agencies, by Activity Category, Fiscal Years 2012 through 2014

Agency	Prevention[a]	Early detection and rapid response[b]	Control and management[c]	Restoration[d]	Research[e]	Education and public awareness[f]	Leadership and international cooperation[g]
Department of Agriculture							
Animal and Plant Health Inspection Service	√	√	√	√	√	√	√
U.S. Forest Service	√	√	√	√	√	√	√
Department of Commerce							
National Oceanic and Atmospheric Administration	√	√	√	√	√	√	√
Department of Defense							
U.S. Army Corps of Engineers	√	√	√	√	√	√	√
Department of Homeland Security							
U.S. Coast Guard	√		√		√	√	√
Department of the Interior							
Bureau of Land Management			√	√	√	√	
Bureau of Reclamation	√	√	√	√	√	√	√
National Park Service	√	√	√	√	√	√	√
U.S. Fish and Wildlife Service	√	√	√	√	√	√	√
U.S. Geological Survey		√			√		√
Department of State	√	√	√	√	√	√	√
Department of Transportation							

Agency	Preventiona	Early detection and rapid responseb	Control and managementc	Restorationd	Researche	Education and public awarenessf	Leadership and international cooperationg
Maritime Administration	√				√		
Environmental Protection Agency	√	√	√	√	√	√	
Total number of agencies conducting activity	11	10	11	10	13	11	10

Sources: GAO questionnaires of 13 Task Force member agencies. I GAO-16-49

Note: These seven activity categories were developed by the National Invasive Species Council Activities and reflect common activities agencies conduct along the continuum of an invasion of a species. These activity categories apply to all invasive species, including both aquatic and terrestrial species.

[a]*Prevention* includes actions taken to prevent the entry, establishment, dispersal, and dissemination of aquatic invasive species.

[b]*Early Detection and Rapid Response* includes actions taken to detect the presence of an aquatic invasive species and assess current and potential impact of the introduction. Rapid Response includes activities taken to eradicate, contain, or control a potentially invasive species introduced into an ecosystem before it spreads.

[c]*Control and Management* includes actions taken to lessen and manage the impact of aquatic invasive species within their established ranges and limit their spread.

[d]*Restoration* includes actions taken to assist the recovery and reestablishment of aquatic plant and animal communities that have been overwhelmed by aquatic invasive species.

[e]*Research* includes actions taken to identify, evaluate, control, and understand aquatic invasive species and their interactions with the biotic and abiotic elements of the environment.

[f]*Education and Public Awareness* includes actions taken to maintain and increase public awareness of aquatic invasive species and programs and to promote public actions that reduce the spread and impact of aquatic invasive species.

[g]*Leadership and International Cooperation* includes actions taken to provide leadership, oversight and coordination to maintain and enhance the capabilities to prevent, control, manage, and understand aquatic invasive species and invasion pathways with relevant state, local, and international partners, and provide for public input and participation.

End Notes

[1] For the purposes of this report, we define an invasive species as a nonnative species—to include all taxa of animals, plants, and microorganisms—the introduction of which does or is likely to cause economic or environmental harm or harm to human health.

[2] GAO, *Invasive Species: Clearer Focus and Greater Commitment Needed to Effectively Manage the Problem*, GAO-03-1 (Washington, D.C.: Oct. 22, 2002); *Invasive Species: Cooperation and Coordination Are Important for Effective Management of Invasive Weeds*, GAO-05-185 (Washington, D.C.: Feb. 25, 2005); CRS, Invasive Non-Native Species: Background and Issues for Congress (Washington, D.C.: Nov. 25, 2002); and the Department of the Interior's *National Invasive Species Council Five-Year Review of Executive Order 13112 on Invasive Species* (Washington, D.C.: 2005).

[3] The Department of the Interior's 2014 Invasive Species Action Plan states that invasive species pose one of the greatest threats to the ecological, economic, and cultural integrity of U.S landscapes. The plan also states that the number and impacts of aquatic invasive species are expected to escalate in the coming decade due to, among other things, the global movement of people and materials from increased tourism and trade that will further disperse species around the world.

[4] EDDMapS. August 2015. The University of Georgia, Center for Invasive Species and Ecosystem Health. Available online at www.eddmaps.org/tools/query.

[5] GAO-05-185; GAO-03-1; and GAO, *Invasive Species: Federal and Selected State Funding to Address Harmful, Nonnative Species*, GAO/RCED-00-219 (Washington, D.C.: Aug. 24, 2000).

[6] D. Pimentel, R. Zuniga, and D. Morrison, "Update on the environmental and economic costs associated with alien-invasive species in the United States," *Ecological Economics* 52 (2005). This study is the most recent comprehensive assessment of the costs associated with invasive species on a national scale available, according to officials from several federal agencies that work on invasive species issues.

[7] 64 Fed. Reg. 6183 (Feb. 8, 1999). National Invasive Species Council, Invasive Species Interagency Crosscut Budget (May 27, 2015). In 1999, Executive Order 13112 established the National Invasive Species Council, to, among other things, provide national leadership regarding invasive species and coordination of federal agency activities concerning invasive species relying to the extent feasible and appropriate on existing organizations. As part of this effort, the National Invasive Species Council has been identifying funding sources and spending by federal agencies on invasive species activities through an annual "crosscut" budget summary.

[8] Expenditures are the actual spending of money; an outlay. GAO, *A Glossary of Terms Used in the Federal Budget Process*, GAO-05-734SP (Washington, D.C.: September 2005).

[9] As we have previously found, a fundamental concept to invasiveness is that invasive species have been introduced into an environment in which they did not evolve, and they usually have no natural predators to limit their spread. GAO-03-1.

[10] 16 U.S.C. § 4701(b).

[11] 16 U.S.C. § 4721(b).The Nonindigenous Aquatic Nuisance Prevention and Control Act of 1990, as amended, designated the following as Task Force members: the Director of the U.S. Fish and Wildlife Service, the Undersecretary of Commerce for Oceans and Atmosphere, the Administrator of the Environmental Protection Agency, the Commandant of the U.S. Coast Guard, the Assistant Secretary of the Army (Civil Works), the Secretary of Agriculture, and the head of any other federal agency the Task Force chairpersons deem

appropriate. As of July 2015, the Task Force was co-chaired by representatives from the U.S. Fish and Wildlife Service and the National Oceanic and Atmospheric Administration, and membership consists of the following 13 federal agencies and departments: (1) Animal and Plant Health Inspection Service, (2) U.S. Forest Service, (3) National Oceanic and Atmospheric Administration, (4) U.S. Army Corps of Engineers, (5) U.S. Coast Guard, (6) Bureau of Land Management, (7) Bureau of Reclamation, (8) National Park Service, (9) U.S. Fish and Wildlife Service, (10) U.S. Geological Survey, (11) Department of State, (12) Department of Transportation's Maritime Administration, and (13) Environmental Protection Agency. In addition, the act, as amended, authorized the chairpersons to invite representatives from state agencies and other governmental entities to participate as ex-officio members of the Task Force. As of 2015, the Task Force included 13 state, regional, and nongovernment entities as ex-officio members, along with six regional panels.

[12] The Task Force uses the terms "aquatic invasive species" and "aquatic nuisance species" interchangeably. The 1990 Act uses the term "aquatic nuisance species." For purposes of this review, we use the term "aquatic invasive species." According to the definition we used in our questionnaire, aquatic species include all animals and plants as well as pathogens or parasites of aquatic animals and plants totally dependent on aquatic ecosystems for at least a portion of their life cycle. Bacteria, viruses, parasites and other pathogens of humans are excluded.

[13] 16 U.S.C. § 4722(b).

[14] 16 U.S.C. § 4722(k)(2).

[15] Pub. L. No. 113-121, § 1039(a)(2), 128 Stat. 1193, 1237 (2014).

[16] We also requested that Task Force member agencies provide us with details about their aquatic invasive species related expenditures, such as expenditures by categories of activities and specific aquatic invasive species of concern. The agencies varied in the level of detail they were able to provide to us. Because of the incompleteness and inconsistency of the data reported across the Task Force member agencies, we did not include this information in our report.

[17] GAO/GGD-96-118.

[18] According to Task Force representatives and FWS officials, in fiscal year 2014, about $260,000 in funding was provided through the FWS' Branch of Aquatic Invasive Species program budget to support administration of the Task Force.

[19] Several agencies have included addressing invasive species (or aquatic invasive species in particular) as part of their strategic planning efforts that guide agency activities and resource expenditures. Specifically, the Department of the Interior, with five member agencies on the Task Force, includes addressing invasive species as a stated goal as part of its departmentwide goals contained in its 2014-2018 strategic plan. Similarly, in its 2015-2020 strategic plan, the U.S. Forest Service includes addressing invasive species as part of its goal to sustain the nation's forests and grasslands. In addition, in September 2015, FWS's Fish and Aquatic Conservation Program released its 2016-2020 strategic plan in which managing aquatic invasive species is identified as one of seven goals. This plan described specific challenges related to aquatic invasive species that program staff plan to conduct over the 5-year period to accomplish the plan's goals.

[20] Each of the six regional panels meets at least once a year to share information and coordinate activities, among other things, and each panel is responsible for reporting back to the Task Force on its progress in implementing aquatic invasive species activities and to provide recommendations. The 1990 Act directed the Task Force to establish two of the six regional panels and to encourage development of additional regional panels.

[21] According to Task Force representatives and FWS officials, in fiscal year 2014 FWS provided over $3 million in funding, including approximately $1 million to support implementation of 40 state and interstate aquatic nuisance species plans with Task Force approved management plans; $240,000 to six regional panels; $1 million for the Quagga-Zebra Mussel Action Plan; $300,000 for the National Asian Carp Management and Control Plan; and $700,000 for expenditures on other national species management and control plans.

[22] David M. Lodge, et al., "Biological Invasions: Recommendations for U.S. Policy and Management" in *Ecological Society of America*, 16(6) (Washington, D.C., 2006); Brian Leung et al., "An ounce of prevention or a pound of cure: bioeconomic risk analysis of invasive species" in *The Royal Society*, DOI 10.1098/rspb.2002.2179; and David Finnoff et al., "Take a risk: Preferring prevention over control of biological invaders," *Ecological Economics* 62 (2007) 216-222; U.S. Environmental Protection Agency, Office of Wetlands, Oceans, and Watersheds, "Overview of EPA Authorities for Natural Resource Managers Developing Aquatic Invasive Species Rapid Response and Management Plans" (Washington, D.C.: December 2005).

[23] David M. Lodge, et al., "Biological Invasions: Recommendations for U.S. Policy and Management" in *Ecological Society of America*, 16(6) (Washington, D.C., 2006).

[24] See also G.M. Ruiz and J. T. Carlton, *Invasive Species: Vectors and Management Strategies*, chapter 5, 2nd ed. (2003); and Daniel Simberloff, "How Much Information on Population Biology Is Needed to Manage Introduced Species?," *Conservation Biology* 17, no. 1 (February 2003). "The most effective way to deal with invasive species, short of keeping them out, is to discover them early and attempt to eradicate or at least contain them before they spread."

[25] Great Lakes Interagency Task Force, *Great Lakes Restoration Initiative Action Plan Fiscal Years 2010-2014*, 09-P-0231 (Washington, D.C.: Feb. 21, 2010.) The plan includes a "zero tolerance policy" for invasive species. We recently reported on information available about the Initiative's activities and results, including those addressing invasive species, see GAO, *Great Lakes Restoration Initiative: Improved Data Collection and Reporting Would Enhance Oversight*, GAO-15-526 (Washington, D.C.: July 21, 2015).

[26] The preference to do more prevention-oriented activities, according to officials from several member agencies, includes activities under both the *prevention* and *early detection and rapid response* activity categories.

[27] To treat aquatic invasive species outside project boundaries in the past, the Corps partnered with state and local agencies. Specifically, the River and Harbor Act of 1958, as amended, authorizes a comprehensive U.S. Army Corps of Engineers program to provide for the prevention, control, and progressive eradication of noxious aquatic plant growths and aquatic invasive species from navigable and other waters of the United States. The law requires a cost share from local interests for the program's projects (known as the Aquatic Plant Control program). Before 1996, the Corps utilized this authority to fund aquatic invasive plant control projects with local and state governments. According to Corps officials, Corps management stopped requesting funding for this control program in 1996 based on a determination that the local and state governments receiving the benefits of this work should be responsible for paying all of the costs. From fiscal year 1998 through fiscal year 2012, the Corps conducted limited activities under this Aquatic Plant Control program when Congress would appropriate or direct funds for it, according to Corps officials.

[28] 78 Fed. Reg. 21938 (Apr. 12, 2013). In October 2015, the United States Court of Appeals for the Second Circuit ruled that EPA acted arbitrarily and capriciously with respect to five

aspects of the permit and remanded these aspects to EPA to adequately address them; however, the permit remains in force while EPA does so.

[29] A species can be added to the list of injurious wildlife by statutory amendment or by FWS rulemaking. The most recent listings were in March 2015, when FWS added four reptiles to the list. See 80 Fed. Reg. 12702 (Mar. 10, 2015). As of April 2015, there were about 240 mammals, birds, reptiles, fish, crustaceans, and mollusks—of which about 150 are considered aquatic invasive species—that were listed as injurious wildlife. This includes species of walking catfish (100), snakeheads (28), zebra mussels (6), carp (4), mitten crab (3) and snakes (9). In addition, FWS officials indicated that all salmonids (approximately 170 species) are listed due to their pathogen risk, which renders them injurious. In October 2015, FWS finalized a change to the process by which it adds new species to its list of injurious wildlife to make it more efficient and allow the agency to better prevent the introduction of species that are injurious. 80 Fed. Reg. 66554 (Oct. 29, 2015).

[30] This work was initiated in response to the October 2014 Priority Agenda Enhancing the Climate Resilience of America's Natural Resources, issued by the White House Council on Climate Preparedness and Resilience.

[31] USGS officials told us that, before 2012, the Nonindigenous Aquatic Species Database also included data on aquatic invasive plants, but that they stopped including plant data in the database because of funding constraints at the time. In June 2015, USGS officials told us that an increase in funding to their Invasive Species Program in fiscal year 2015 has allowed them to resume their efforts to collect and report data on aquatic invasive plants as part of this database, and that adding plant information back to the database was a high funding priority for the agency partly due to requests from other federal agencies and partners to do so.

[32] 33 U.S.C. § 610(a)(1).

[33] Some regional panel representatives told us that it has been difficult for them to fully conduct activities, including attending semiannual Task Force meetings, due to decreased funding provided to the panels. Beginning in fiscal year 2012, funding for the regional panels decreased from $300,000 to $240,000 annually, which represents a decrease from $50,000 to $40,000 per panel. Similarly, Task Force representatives said that funding to support implementation of state and interstate management plans has remained level at about $1 million annually, but that the number of approved plans eligible for assistance has increased each year, resulting in less funding being awarded per state plan.

[34] For the Task Force's 2013-2017 strategic plan, see http://anstaskforce.gov/Documents/ANSTF%20Strategic%20Plan%202013-2017.pdf.

[35] We have previously found that, though a legal requirement specific to federal agencies and departments, the Government Performance and Results Modernization Act of 2010 serves to guide executive federal agency planning and provides leading practices for strategic planning at agencies and entities that work closely with or are a part of the federal government, including task forces. (See GAO, *Environmental Justice: EPA Needs to Take Additional Actions to Help Ensure Effective Implementation*, GAO-12-77 (Washington, D.C.: Oct. 6, 2011.) The identification of goals in the Task Force's Strategic Plan is consistent with such leading practices.

[36] We have previously found that the Government Performance and Results Modernization Act of 2010—which calls for a performance plan to implement an agency's strategic plan—provides leading practices for strategic planning for task forces. The operational plan described in the Task Force's Strategic Plan is consistent with such leading practices.

[37] The 1990 Act requires the Task Force to submit a report to Congress annually detailing the progress of its program, as well as to develop a program that (1) describes the specific prevention, monitoring, control, education and research activities to be conducted by each Task Force member; (2) describes the role of each Task Force member in implementing the elements of the program; and (3) includes recommendations for funding to implement elements of the program.

[38] All six regional panels also provided information to the Task Force for the reporting matrix in 2012.

[39] As a result of sequestration in 2013 and 2014, federal agencies reduced or delayed some services and disrupted agency operations. For information on 2013 sequestration, see GAO, *2013 Sequestration: Agencies Reduced Some Services and Investments, While Taking Certain Actions to Mitigate Effects*, GAO-14-244 (Washington, D.C.: Mar. 6, 2014).

[40] The 1990 Act requires the Task Force, beginning in 1992, to submit a report to Congress annually detailing the progress of its program.16 U.S.C. § 4722(k)(2).

[41] GAO, *Climate Change: Adaptation: Strategic Federal Planning Could Help Government Officials Make More Informed Decisions*, GAO-10-113 (Washington, D.C.: Oct. 7, 2009).

End Notes for Appendix I

[1] 16 U.S.C. §§ 4701-4751.

[2] We defined our scope to include the 13 federal department and agencies that are the federal members of the Aquatic Nuisance Species Task Force, as established by the 1990 Act. Other federal agencies may also conduct aquatic invasive species-related activities.

[3] We also included an eighth category labeled "other" for member agencies to report any activities they determined did not fit within any of the seven categories.

[4] 50 C.F.R. pt. 80-86.

[5] 16 U.S.C. § 4722(b).

[6] GPRA Modernization Act of 2010, Pub. L. No. 111-352, 124 Stat. 3866 (2011).

[7] GAO/GGD-96-118.

In: Aquatic Invasive Species
Editor: Ernestine Sandoval

ISBN: 978-1-63485-391-0
© 2016 Nova Science Publishers, Inc.

Chapter 2

AQUATIC NUISANCE SPECIES TASK FORCE STRATEGIC PLAN (2013-2017)*

Aquatic Nuisance Species Task Force

EXECUTIVE SUMMARY

Aquatic nuisance species (ANS) are nonindigenous species that threaten the diversity or abundance of native species, the ecological stability of infested waters, and/or any commercial, agricultural, aquacultural, or recreational activities dependent on such waters. ANS include nonindigenous species that may occur within inland, estuarine, or marine waters and that presently or potentially threaten ecological processes or natural resources. The term ANS is often used interchangeably with aquatic invasive species, the preferred term of Federal and State managers. An aquatic invasive species is defined as a species not native to the ecosystem under consideration whereby introduction of this species does or is likely to cause economic or environmental harm or threaten human health.

ANS are one of the largest threats to the ecosystems and economies of the United States. Approximately 49% of the species on the threatened or endangered species lists are at risk primarily because of predation or competition with exotic species. In fact, impacts of invasive species are second only to habitat destruction as a cause of global biodiversity loss. ANS such as

---

* This is an edited, reformatted and augmented version of a document approved by the ANS Task Force, May 3, 2012.

snakehead fish, sea lamprey, hydrilla, and the New Zealand mudsnail may prey upon, displace, alter habitat, or otherwise harm native species. Other ANS may reduce production of fisheries, decrease water availability to residential and commercial users, block transport routes, choke irrigation canals, foul industrial and public water supply pipelines, degrade water quality, accelerate filling of lakes and reservoirs, and decrease property values. The damages to human enterprises caused by ANS results in enormous economic costs. The United States invests more than $120 billion per year in damage and control costs to combat invasive species. As the world trade network continues to grow, the number and frequency of introduced species are expected to increase. Additionally, climate change may also allow increased introductions. This plan presents the strategic priorities designed to address ANS in the United States under current and future conditions.

In 1990, Congress passed the Nonindigenous Aquatic Nuisance Prevention and Control Act to establish a broad national program to prevent introduction and control the spread of introduced ANS; this legislation was reauthorized and furthered with the National Invasive Species Act in 1996. The Aquatic Nuisance Species Task Force (ANSTF) is an intergovernmental organization dedicated to preventing and controlling aquatic invasive species and implementing these Acts. The ANSTF, co-chaired by the U.S. Fish and Wildlife Service and National Oceanic and Atmospheric Administration, consists of 13 Federal agency representatives and 13 *ex-officio* representatives. These members work in conjunction with Regional Panels and issue-specific committees to coordinate efforts amongst agencies as well as efforts of the private sector and other North American interests.

The ANSTF serves to develop and implement a program for waters of the United States to prevent introduction and dispersal of ANS, monitor, control, and study such species, and disseminate related information. In 1994, the *Aquatic Nuisance Species Program* document was drafted to guide the work of the ANSTF, establishing the core elements of the Task Force. A broader focus was established with the 2002-2007 and 2007-2012 Strategic Plans, placing more emphasis on prevention strategies. This *Aquatic Nuisance Species Task Force Strategic Plan 2013- 2017* (Strategic Plan) carries through many of the goals and objectives established in previous plans by remaining focused on prevention, monitoring, and control of ANS as well as increasing public understanding of the problems and impacts associated with invasive species. The Strategic Plan also calls attention to other areas of management including habitat restoration and research. The Strategic Plan establishes eight goals, each with objectives and action items to be completed in the next 5 years.

Coordination: The ANSTF was created to facilitate cooperation and coordinate efforts between Federal, State, tribes, and local agencies, the private sector, and other North American interests. The objectives for the coordination goal include strengthening cooperation at both national and regional levels within the ANSTF and the Regional Panels and encouraging the development and implementation of ANS plans and regulations.

Prevention: Prevention is the first-line of defense against ANS. This goal calls for developing strategies to identify and reduce the risk of ANS introduced by increasing development and use of risk assessments, Hazard Analysis and Critical Control Point programs (HACCP), and pathway assessment and interdiction options.

Early Detection and Rapid Response: Early Detection and Rapid Response programs are designed to monitor habitats to discover new species soon after introduction, report sightings of previously unknown species in an area, and work quickly to keep the species from becoming established and spreading. Objectives for the ANSTF include improving detection and monitoring programs and facilitating development and implementation of rapid response contingency plans.

Control and Management: Control and management tools are needed to assess, remove, and contain ANS populations as well as to guide management decisions. The ANSTF will implement this goal by evaluating and providing support to management plans, increasing training opportunities, and encouraging the development of management techniques.

Restoration: Habitat restoration is an essential to guard against future invasions and to minimize harm from invasive species. This goal focuses on restoring impacted ecosystems and consideration of potential ANS during planning and implementation of restoration activities.

Education/Outreach: The lack of awareness concerning ANS impacts is one of the largest management obstacles. Few people understand the threat some nonindigenous species pose and how their actions might introduce them. Objectives by the ANSTF for education and outreach include reaching out to the general public, providing technical guidance to targeted audiences, and raising awareness among legislators and decision makers.

Research: Research supports all facets of the Strategic Plan and is necessary to increase the effectiveness of prevention, detection, response, and control and management of invasive species. To help ensure that research addresses critical needs, this goal focuses on coordination among government agencies, academia, and other participating entities.

Funding: Securing dedicated long-term and emergency funding is necessary to achieve the goals laid out in the Strategic Plan. The actions outlined by the ANSTF focus on coordinating Federal agency budgets to support ANSTF priorities, develop partnerships, and seek opportunities to leverage funds within Federal and State agencies, local governments, tribal entities, industry, as well as other entities including non-governmental organizations.

The Strategic Plan should not be considered a comprehensive list of all ANS strategic actions; it does contain a targeted set of priority strategic goals, objectives, and associated action items that are intended to be completed in the next 5 years. The accomplishment of specific objectives and action items will be dependent upon budgets of individual agencies and the Regional Panels; and in some cases, legal or regulatory changes as well as enforcement of these changes. An Operational Plan will be composed to depict short-term efforts to achieve the actions in the Strategic Plan to ensure the goals and objectives of the Strategic Plan are measurable and accountable. The Operational Plan will be completed by the ANSTF members working together and separately with support of the Regional Panels and committees. The actions in the Operational Plan will be updated regularly and reported on to measure the progress towards meeting the goals of the Strategic Plan.

Management of ANS is challenging; however, considerable success is being achieved in the prevention, detection, eradication, control, and outreach efforts of ANS along with increased emphasis for the restoration of ecosystems that have been impacted by ANS. Additional research and information exchange, new detection and eradication techniques, and innovative control methodologies are increasing our capacity to address invasive species problems. The *Aquatic Nuisance Species Task Force Strategic Plan 2013–2017* takes a deliberate, cooperative approach and builds on existing programs. The Task Force will strive to maximize efforts over the next 5 years to prevent and control invasive species with the purpose of protecting our environment, economy and human health.

INTRODUCTION

Aquatic nuisance species (ANS) are nonindigenous species that threaten the diversity or abundance of native species, the ecological stability of infested waters, and/or any commercial, agricultural, aquacultural, or recreational activities dependent on such waters. ANS include nonindigenous species that may occur within inland, estuarine, or marine waters and that presently or potentially threaten ecological processes or natural resources. In addition to the severe and permanent damage to the habitats they invade, ANS may also adversely impact society by hindering economic development, preventing recreational and commercial activities, decreasing the aesthetic value of nature, and serving as vectors[1] of human disease. The table below provides a list of the three classes of adverse impacts caused by ANS.

The term ANS is often used interchangeably with aquatic invasive species (AIS), the preferred term of Federal and State managers. An aquatic invasive species is defined as a species not native to the ecosystem under consideration whereby introduction of this species does or is likely to cause economic or environmental harm or threaten human health.

Selected Examples of Aquatic Nuisance Species Impacts		
Environmental Effects	Economic Impacts	Public Health
Habitat Alterations	Industrial Water Supplies	Disease Epidemics
Water Quality	Municipal Water Supplies	Public Safety
Predation	Power Plants	Physical Injury
Competition	Commercial Fisheries	Bacterial Risks
Hybridization	Recreation	Harmful Algal Blooms
Parasitism	Navigation and Shipping	Parasites
Introduction of Pathogens	Aquaculture	Flooding

Environmental Harm

In the United States, approximately 49% of the species on the threatened or endangered species lists are at risk primarily because of predation or competition with exotic species.[2] In fact, impacts of invasive species are second only to habitat destruction as a cause of global biodiversity loss.[3] ANS impact the habitats they invade by reducing the abundance of native species and altering ecosystem processes. They can impact native species through predation, competition for food and space, hybridization, as well as the introduction of pathogens and parasites. Normal functioning of the ecosystem, including hydrology, nutrient cycling, or productivity, may also be altered by ANS. Aquatic weeds provide an excellent example of the severe impact an exotic organism may have on the environment. Non-native aquatic plants including Eurasian water-milfoil (*Myriophyllum spicatum*), hydrilla (*Hydrilla verticillata*), and water hyacinth (*Eichhornia crassipes*) can negatively impact the diversity of native aquatic plants and invertebrates, the efficiency for large predator fish to obtain prey, water quality, and aquatic recreational activities including swimming, fishing and boating. Invasive fish species also have the ability to alter aquatic ecosystems. For example, the common carp (*Cypinus carpio*) is capable of reducing native vegetation and increasing turbidity. These types of habitat alterations are responsible for the extinction of several native fish species.[4]

Economic Harm

ANS are seen as a threat not only to native biodiversity and ecosystem functioning, but also to economic development. They can reduce production of fisheries, decrease water availability to residential and commercial users, block transport routes, choke irrigation canals, foul industrial and public water supply pipelines, degrade water quality, accelerate filling of lakes and reservoirs, and decrease property values. The damages to human enterprises caused by ANS result in enormous economic costs.

Over the past 200 years, more than 50,000 non-native plant and animal species have become established in the United States. Approximately one in seven has become invasive,[5] with damage and control costs estimated at more than $120 billion per year[6] - a cost higher than the total of all other natural disasters combined.[7] Zebra and quagga mussels (*Dreissena polymorpha, D. rostriformis bugensis*) alone cause one billion dollars per year in damages.[8] Another 100 million is spent annually in the United States to control nonnative aquatic weeds.[9] In two California lagoons, more than $5 million was spent in the first 3 years of an ongoing eradication program for the seaweed *Caulerpa taxifolia*.[10] As a final example, the Great Lakes States invested over $26.7 million toward prevention and control of aquatic invasive species in just 2 years, of which almost $900,000 was committed to Asian carp[11] control efforts.[12] These numbers are likely underestimated as they do not consider ecosystem health or the aesthetic value of nature, which can influence tourism and recreational revenue. Estimating the economic impact associated with ANS is further confounded as monetary values are difficult to estimate for the extinction of species or loss of native biodiversity and ecosystem services.

Harm to Human Health

Throughout history, epidemic diseases such as malaria, yellow fever, typhus, and bubonic plague have spread using organisms as vectors and reservoirs. Further, there has been conjecture that the ballast water of ships may transport waterborne pathogens and diseases[13] as well as causative agents of harmful algal blooms (HABs),[14,15] however, additional study is needed to link the transport of microorganims to outbreaks of human disease. The effect of ANS on public health extends beyond the immediate effects of disease and parasites; human injury may also result from ANS. For instance, hazards may occur from collisions between boaters and jumping silver carp (*Hypophthalmichthys molitrix) or* from the sharp-edged mussel shells found in recreational areas. Additional risk to human is perceivable as chemicals used to control invasive species can pollute soil and water. Other ANS, such as invasive mussels, may increase human and wildlife exposure to organic pollutants such as polychlorinated biphenyls (PCBs), as these toxicants accumulate in their tissues and are passed up the food chain.

Ans - What Can Be Done?

Prevention is the most cost effective and environmentally protective tool to control ANS. Preventative measures include decontaminating boats and gear that could transport ANS and restricting the importation or release of potentially harmful species. However, even the best prevention efforts may not stop all invasions. When a new species is introduced, the best strategy is early detection and rapid response. This includes monitoring habitats to discover new species soon after introduction, reporting sightings of previously unknown species in an area, and working quickly to keep introduced species from becoming established and spreading. Once established, invasive species can be difficult to control and nearly impossible to eradicate. Control methods include mechanical, chemical, and biological approaches. The methods used will vary dependent upon the species and location; however, control efforts are typically costly, labor intensive, and indirectly affect native species. Furthermore, control efforts often create disturbance, which may render the habitat vulnerable to subsequent invasions. Accordingly, habitat restoration is necessary following eradication or control efforts to minimize the chance an area will be reinvaded.

Education and outreach are critical tools to prevent and manage the impacts of ANS. The public must understand the problems and impacts associated with ANS so they can be active partners in solving the problem. More importantly, people need to know what they can do to help prevent the introduction and spread of ANS. Federal, State, and local programs and legislation have implemented regulations to prevent the introduction and reduce the spread of invasive species. These regulations include mandatory boat inspections at public boat ramps, live species prohibitions and restrictions, and ballast water regulations at shipping ports. To successfully address ANS issues collaboration, cooperation, and coordination are necessary among and between Federal and State agencies, local governments, tribal entities, industry, as well as other entities including non-governmental organizations.

Future Challenges

Global trade and intercontinental travel have been cited as major causes of biological invasion. For example, it has been estimated that 10,000 marine species are transported around the world in ballast water every day.[16] As the world trade network continues to grow, new markets and trade routes continually open. This growth will increase the number of new species that are introduced and the frequency with which such introductions are repeated. Managing this increased rate of aquatic bioinvasion will require the United States and other counties to strengthen approaches for preventing introductions while maintaining trade.

Additional challenges to ANS management result from changes in the Earth's climate that will likely continue, or even accelerate, over the next century. Very little is known of the impacts from ANS in relation to climate change, yet models suggest that the economic, energy, social, and environmental impacts may be profound. Fast growth, rapid reproduction, and the ability to survive in a wide range of environmental conditions are among some of the life history traits shared by ANS that may allow them to capitalize on the biotic and abiotic changes generated by global climate change. Furthermore, species that have long been "in motion," but failed to establish and reproduce in hostile conditions, may soon be able to invade these once "off limit" thermal regimes. Other species will migrate to maintain the temperature conditions needed for reproduction, growth, and feeding. There is a growing concern that these shifting species will begin to function as invasive species, disrupting the structure and function of their new communities. Many communities have already experienced the impacts of warming coastal waters and have shown subsequent alterations in species proportions as well as changes to community structure and dynamics. Predictions of how species will respond to climate change will help guide conservation decisions and management of natural resources. Future ANS managers will need to develop tools that include both invasion biology and climate change impacts.

STRUCTURE OF THE AQUATIC NUISANCE SPECIES TASK FORCE

The Aquatic Nuisance Species Task Force (ANSTF) was established by Congress with the passage of the Nonindigenous Aquatic Nuisance Prevention and Control Act (NANPCA) in 1990 and reauthorized with the passage of the National Invasive Species Act (NISA) in 1996 (collectively, the Act). The ANSTF is an interagency committee established by Section 1201 of the Act and serves to develop and implement a program for waters of the United States[17] that:

- Prevents the introduction and dispersal of ANS;
- Monitors, controls and studies such species;
- Conducts research on methods to monitor, manage, control and/or eradicate such species;
- Coordinates ANS programs and activities of ANSTF members and affected State agencies; and
- Educates and informs the general public and program stakeholders about the prevention, management, and control of these species

Federal and Ex-Officio Members

The ANSTF's charter is authorized by the Federal Advisory Committee Act (FACA) of 1972. The charter provides the ANSTF with its core structure and ensures an open and public forum for its activities. To meet the challenges of developing and implementing a coordinated and complementary Federal program for ANS activities, the ANSTF members include 13 Federal agency representatives and 13 representatives from ex-officio member organizations. These members work in conjunction with Regional Panels and issue-specific committees to coordinate efforts amongst agencies as well as efforts of the private sector and other North American interests.

The Act designated the Director of the Fish and Wildlife Service and the Undersecretary of Commerce for Oceans and Atmosphere as the ANSTF Co-chairpersons. It specified six Federal agencies[18] that would constitute the ANSTF, but also gave the co-chairs legal authority to include other Federal agencies as members of the ANSTF, as appropriate. Members of ANSTF are responsible for committing resources to achieve the goals of the Strategic Plan and reporting annually on their progress. At the time of Plan adoption (May 3, 2012), the following were ANSTF member departments and agencies:

- United States Fish and Wildlife Service (USFWS)—co-chair
- National Oceanic and Atmospheric Administration (NOAA)—co-chair
- Army Corps of Engineers (ACOE)
- Bureau of Land Management (BLM)
- Bureau of Reclamation (BOR)
- Department of State (DOS)
- Environmental Protection Agency (EPA)
- United States Forest Service (USFS)
- Department of Transportation (DOT), Maritime Administration (MARAD)
- National Park Service (NPS)
- United States Coast Guard (USCG)
- United States Department of Agriculture, Animal and Plant Health Inspection Service (USDA-APHIS)
- United States Geological Survey (USGS)

The Act also named four organizations[19] to be included in the base ANSTF membership, yet authorized the co-chairs to invite representatives of specific regional organizations, State agencies, and other governmental entities to participate as *ex-officio* members of the ANSTF. At the time of Plan adoption (May 3, 2012), the following were ex-*officio* members of the ANSTF:

- Great Lakes Commission
- Lake Champlain Basin Program
- Chesapeake Bay Program
- San Francisco Estuary Project
- American Public Power Association
- American Water Works Association
- Association of Fish and Wildlife Agencies
- Gulf States Marine Fisheries Commission
- Mississippi Interstate Cooperative Resources Association
- Native American Fish and Wildlife Society[20]

- National Association of State Aquaculture Coordinators
- Smithsonian Environmental Research Center
- *Fisheries and Oceans Canada (invited observer)*

ANSTF Regional Panels

The ANSTF focuses its work on ANS issues of national concern that require or could benefit from collaborative solutions. It strives to create opportunities and synergies among members and participants to work collaboratively by sharing resources, expertise, and ideas across agency and organizational lines. While the ANSTF has a national focus, it recognizes the tremendous importance of actions taken at the regional and local level to achieve national ANS solutions. Section 1203 of NANPCA created the Great Lakes Regional Panel to identify priorities, to coordinate ANS program activities, and to advise public and private interests on control efforts in their region. The 1996 amendment required the ANSTF to encourage the development of additional regional panels to provide an intergovernmental mechanism for the development of a coordinated Federal program to prevent and control nonindigenous ANS as authorized by the Act. The Regional Panels are responsible for implementing actions that assist in achieving the Strategic Plan's goals and reporting annually on their progress.[21] At the time of Plan adoption (May 3, 2012), the ANSTF had established six Regional Panels:

- Great Lakes Regional Panel
- Western Regional Panel
- Gulf and South Atlantic Regional Panel
- Northeast Regional Panel
- Mississippi River Basin Regional Panel
- Mid-Atlantic Regional Panel

Regional Panel membership is composed of Federal, State, intergovernmental, tribal, industry, and environmental non-governmental organization representatives. In addition, some panels may include international representation (e.g., Canadian Federal and Provincial representatives) as observers.

The Regional Panels make a concerted effort to involve a broad spectrum of stakeholders in order to provide balanced advice to the ANSTF on regional priorities and issues of local significance.

ANSTF Committees

To obtain the necessary technical coordination of various ANS efforts, the ANSTF has established several issue- and/or species-specific standing and ad hoc committees to carry out the Act. Committee membership includes all affected entities, such as Federal and State agencies, tribes, non-governmental organizations, industry groups, and academia. Committees are made up of member agency representatives and subject matter experts. Committee activities include the development of public awareness/action campaigns, species-specific control and management plans, standardized scientific protocols, research priorities, theoretical frameworks to screen organisms prior to their entry to the United States., and providing technical advice to the ANSTF. Committees, both standing and ad hoc, are responsible for reporting to the ANSTF on goal attainment under this Plan. At the time of Plan adoption (May 3, 2012), the ANSTF had established three standing committees that oversee related working groups: the Communication, Education and Outreach Committee, Research Committee, and Prevention Committee (the latter is a joint committee with the National Invasive Species Council (NISC)). In addition, ad hoc committees are formed as needed to focus on a specific discipline or issue that warrants the attention of the ANSTF.

ANSTF Strategic Plan

Section 1202 of the Act authorized the ANSTF to develop and implement a program for waters of the United States to prevent introduction and dispersal of ANS, to monitor, control, and study such species, and to disseminate related information. The *Aquatic Nuisance Species Program* document guided the work of the ANSTF from 1994 to 2002. The document tracked the requirements outlined in the NANPCA (1990). It established the core elements of the ANS program (prevention, detection and monitoring, control) and support elements (research, education, and technical assistance), provided for prioritization of activities, and charted a course for implementation of the Act.

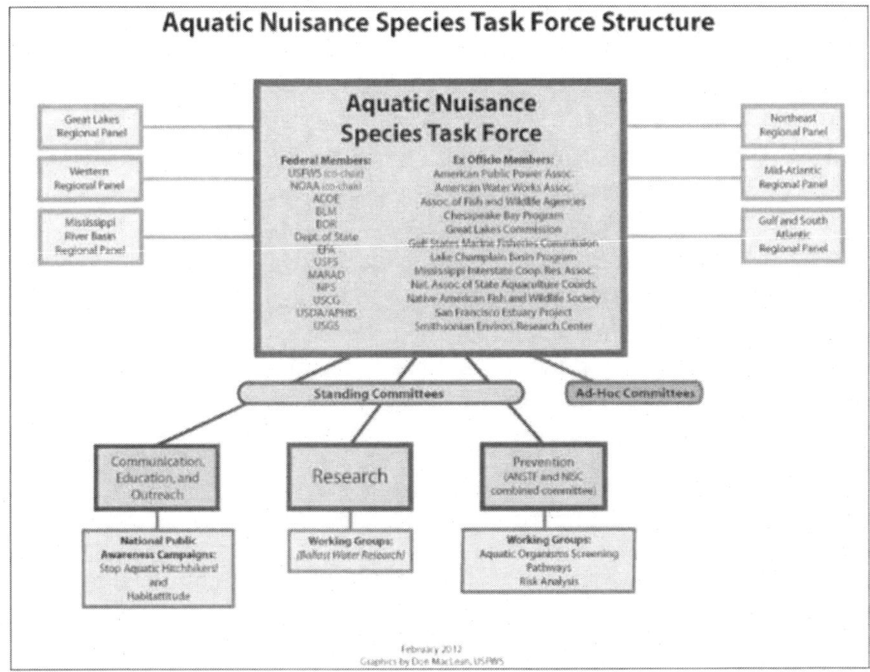

The ANSTF Strategic Plans for 2002–2007 and 2007- 2012 maintained the key elements of the ANS Program, but provided a broader focus for activities consistent with provisions of NISA (1996). These plans provided more emphasis on prevention strategies, particularly for intentional introductions.

The ANSTF Strategic Plan for 2013 – 2017 (hereafter, the Strategic Plan) carries through many of the goals and objectives established in previous plans by remaining focused on prevention, monitoring, and control of ANS as well as increasing public understanding of the problems and impacts associated with invasive species. The Strategic Plan also calls attention to other areas of ANS management, including habitat restoration and research. The Strategic Plan establishes eight goals:

1. Coordination - Maximize the organizational effectiveness of the Aquatic Nuisance Species Task Force
2. Prevention - Develop strategies to identify and prevent the establishment of new and slow the spread of existing ANS in the waters of the United States

3. Early Detection and Rapid Response - Identify and respond to aquatic nuisance species within a timely manner following introduction in order to prevent their establishment and/or spread
4. Control and Management - Control established aquatic nuisance species when feasible and when the benefits of managing the established species outweigh the costs of removing them with respect to harm to the environment, the economy, and public health
5. Restoration – Protect and rehabilitate native species and ecosystems by conducting habitat restoration efforts on multiple scales
6. Education / Outreach - Increase awareness concerning the threats of aquatic nuisance species, emphasizing the impacts, importance of prevention and containment, and recommendations for appropriate domestic and international actions
7. Research - Facilitate research to address environmental, economic, and human health risks and impacts associated with aquatic nuisance species
8. Funding - Coordinate Federal agency budgets to support Aquatic Nuisance Species Task Force priorities and establish a clear process that links State and regional needs in their areas of responsibility

The strategic goals serve as a blueprint and coordination tool for the ANSTF. The order in which the goals are presented in the Plan represent the logical arrangement determined by the ANSTF; accordingly, the hierarchy of the goals do not represent individual importance or priority level. Under each goal, objectives describe what is to be accomplished over the next 5 years. Action items listed under the objectives describe how ANSTF expects to accomplish the goals and objectives. The accomplishment of specific objectives and action items will be dependent upon budgets of individual agencies and the Regional Panels; and in some cases, legal or regulatory changes as well as enforcement of these changes. The Strategic Plan should not be considered a comprehensive list of all ANS strategic actions; it does contain a targeted set of priority strategic goals, objectives, and associated action items that are intended to be completed in the next 5 years.

ANSTF Operational Plan

In addition to its Strategic Plan, the ANSTF will compose a separate Operational Plan. Its function is to ensure the goals and objectives of the

Strategic Plan are measurable and accountable. In contrast to the action items in the Strategic Plan that outline long-term or continual actions, the actions listed in the Operational Plan will depict short- term efforts used to support and implement the Strategic Plan. The Operational Plan will be completed by the ANSTF members working together and separately with support of the Regional Panels and committees. To the greatest extent possible, implementation of the actions within the Operational Plan will build on and fill gaps in existing activities and programs rather than supplanting them. Responsibilities will be assigned to specific agencies, Regional Panels, or committees. Implementation will be assumed by those specified in line with their specific mandates, priorities, expertise, and funding. The Operational Plan will include, when available, the time frame, lead and supporting agencies or groups, and allocated funding. Further, the Operational Plan will be regularly amended to reflect changes in circumstances, plans, or priorities. The actions listed in the Operational Plan will be reported annually to measure the progress towards accomplishing the over-arching goals and objectives identified within the Strategic Plan.

COORDINATION WITH OTHER FEDERAL AND STATE INVASIVE SPECIES MANAGEMENT PLANS

The ANSTF recognizes that many Federal and State agencies, interagency groups, and local entities contribute to the management of invasive species. The largest and most comprehensive invasive species-focused committees and working groups are the National Invasive Species Council (NISC), Federal Interagency Committee for the Management of Noxious and Exotic Weeds (FICMNEW), and the Federal Interagency Committee on Invasive Terrestrial Animals and Pathogens (ITAP). As described below, these working groups facilitate communication and collaboration at all levels of the Federal government and with State, local and private partners by focusing on particular taxa and pathways.

The National Invasive Species Council (NISC) was established by Executive Order 13112 (Order). NISC is co-chaired by the Secretaries of Agriculture, Commerce, and the Interior and includes 10 member departments and their constituent agencies as well as a small staff assigned specifically to the Council. The Order directs the Secretary of the Interior to establish an Invasive Species Advisory Committee (ISAC) composed of nonfederal experts and stakeholders to provide advice and recommendations to NISC on invasive species-related issues. NISC provides national leadership and oversight on both terrestrial and aquatic invasive species and ensures that Federal programs and activities to prevent and control invasive species are coordinated, effective, and efficient. NISC has specific responsibilities including promoting action at State, tribal, local, and ecosystem levels; identifying recommendations for international cooperation; facilitating a coordinated network to document, evaluate, monitor invasive species' effects; developing a web-based information network on invasive species; and developing guidance on invasive species for Federal agencies to use in implementing the National Environmental Policy Act. NISC is also responsible for preparing a National Invasive Species Management Plan which directs Federal efforts to prevent, control and minimize invasive species and their impacts.

The Federal Interagency Committee for the Management of Noxious and Exotic Weeds (FICMNEW) was established through a Memorandum of Understanding signed by agency leadership in August 1994. FICMNEW represents an unprecedented formal partnership between 16 Federal agencies with direct invasive plant management and regulatory responsibilities spanning across the United States and territories. FICMNEW members interact on important national and regional invasive plant issues and share information with various public and private organizations participating with the Federal sector to address invasive plant issues. FICMNEW's charter directs the committee to coordinate, through the respective Secretaries, Assistant Secretaries, and Agency heads, information regarding the identification and extent of invasive plants in the United States and to coordinate Federal agency management of these species. FICMNEW accomplishes these portions of its charter by developing and sharing scientific and technical information, fostering collaborative efforts among Federal agencies, providing recommendations for national and regional level management of invasive plants, and sponsoring technical and educational conferences and workshops concerning invasive plants.

ITAP is the Federal Interagency Committee on Invasive Terrestrial Animals and Pathogens. ITAP's mission is to support and facilitate more efficient networking and sharing of technical information for program planning and coordination among Federal agencies and departments involved with invasive species research and management. ITAP focuses on several major taxonomic groups of invasive species for which improved technical coordination is essential to facilitate effective Federal responses.

Invasive species management plans prepared by these groups and others provide an opportunity to identify priorities and establish cooperative, well-coordinated approaches to invasive species management. Invasive species issues are significant in their breadth and scope; as this area involves all taxa of life and threatens natural ecosystems around the globe. A variety of pathways are capable of transporting species into new environments including ballast water and hulls of ships, materials associated with the trade of nursery stock, importation of fruits and vegetables, and the international movement of people. Furthermore, multiple efforts are necessary to encompass the various components of invasive species management, including prevention, monitoring, removal, restoration, research, and education. The great extent needed to manage invasive species and their impacts infers that one agency, task force, or work group cannot tackle it alone. Work can be done with greater effectiveness and efficiency if it is focused on specific taxonomic, ecosystem, or regional priorities. To ensure that redundancy and overlap does not occur while addressing invasive species-related issues, the ANSTF communicates with Federal and State agencies and interagency groups to create a big picture framework for existing management plans. Working with other agencies and organizations allows the ANSTF to identify areas where legislation is needed to fill gaps in statutory authority, suggest priority policy issues, and define roles and responsibilities for managing invasive species.

States also play a critical role in preventing and controlling the spread of invasive species and have numerous programs relating to the wide variety of invasive species found within their borders. ANSTF encourages State and interstate planning entities to develop management plans describing detection and monitoring efforts of ANS, prevention efforts to stop their introduction and spread, and control efforts to reduce their impacts. In addition, these plans serve to coordinate efforts between State agencies, local governments, tribal entities, industry, as well as other entities including nongovernmental organizations. Consequently, they are a valuable and effective tool for identifying and addressing ANS problems and concerns in a climate of many jurisdictions and other interested entities. Once a State or interstate ANS management plan is approved, the ANSTF monitors the activities of the planning entity to ensure the plans are implemented. This monitoring process allows the ANSTF to evaluate the capacity and capability at State and local levels to coordinate, detect, and respond to invasive species. This information allows the ANSTF to better identify strategies for monitoring, containment, outreach, and other ANS activities. It also allows for identification of priority activities and species, obstacles to fully implementing the ANS State management plans, and cooperative partnerships that exist among entities managing ANS. This information is critical for recognizing the amount and type of data and management methods available, which allows for an assessment of gaps, redundancies, and opportunities for collaboration among agencies that are not being realized. It is clear that actions and goals performed at the Federal level will not will not succeed unless they are undertaken in cooperation with stakeholders; for that reason, coordination and joint action with State partners is critical for addressing invasive species problems within the United States.

CONCLUSION

Management of ANS is challenging; however, considerable success is being achieved in the prevention, detection, eradication, control, and outreach efforts of ANS along with increased emphasis on the restoration of ecosystems that have been impacted by ANS. Additional research and information exchange, new detection and eradication techniques, innovative control methodologies, and collaborative models are increasing our capacity to manage ANS. Since the establishment of the ANSTF, awareness of the problems caused by ANS has dramatically improved, as evidenced by increased activity at Federal, State, and local levels. Despite the significant increase in activity and awareness, much remains to be done to prevent and mitigate the impacts of ANS. The intent of the ANSTF Strategic and Operational Plans is to create a strategic approach to minimize harm to the environment, economy, or human health that results from ANS. This involves taking advantage of what has been learned and creating next steps that are well planned and coordinated.

Goal 1: Coordination - Maximize the Organizational Effectiveness of the Aquatic Nuisance Species Task Force

The scope and complexity of ANS management summons the strengths of different government agencies and private organizations in different ways. A primary objective of the ANSTF is to facilitate cooperation and coordinate Federal government efforts relating to ANS in the United States with those of the private sector and other North American interests by utilizing Regional Panels and issue-specific committees and including bi-national bodies where applicable. The Regional Panels established by the ANSTF are a critical and effective mechanism for achieving the goals and objectives established by the ANSTF Strategic Plan. The memberships within each of the panels include representatives of States, Indian tribes, non-governmental organizations, commercial interests, and neighboring countries. The roles of each panel include, but are not limited to, identifying regional ANS priorities, coordinating ANS program activities in the region, making recommendations to the ANSTF, and providing advice to public and private interests concerning methods of ANS management and control.

To further increase coordination, the ANSTF encourages State and interstate planning entities to develop management plans describing detection and monitoring efforts of ANS, prevention efforts to avert their establishment and spread, and control efforts to reduce their impacts. In addition, management and control plans have been developed or are under development by the ANSTF and other partners for several species. Under Section 1204 of the Act, management plan approval from the ANSTF is required for both State and species plans to obtain funding. Regardless of financial incentives, plans are a valuable and effective tool for identifying and addressing ANS problems and concerns in a climate of many jurisdictions and interested entities.

Fulfilling Goal 1 requires ongoing cooperation, communication, and dialogue as well as an understanding of the views and roles of all agencies and organizations involved. The actions suggested below will allow the ANSTF to lower institutional barriers to efficiency and effectiveness, beginning with enhanced Federal agency collaboration. The actions also include a thorough analysis of the roles, responsibilities, and supporting legislation involving ANS so gaps in authority can be identified. Further, strong monitoring and evaluation of the ANSTF Strategic and Operational Plans are encouraged to provide measures of success toward reaching goals and providing information for future revisions of the plans.

Objective 1.1: Strengthen the Coordination Capacity of the ANSTF

a) Increase communication among the members, Regional Panels, and committees of the ANSTF to prioritize issues and activities

b) Continue to build capacity within and among the Regional Panels in order to increase regional and international coordination, identify needs and emerging issues, and communicate recommendations to ANSTF members

c) Provide technical guidance and resource assistance to States though a coordinated effort by the Regional Panels

d) Strengthen the working relationship with NISC, working cooperatively to implement ANS activities identified in the ANSTF Strategic Plan and NISC National Management Plan

e) Identify, increase communication, and encourage participation with other interagency organizations dedicated to invasive and nuisance species

f) Identify, increase communication, and encourage participation with industry, tribes, and other stakeholders affected by ANSTF and Regional Panel activities

g) Annually report on the ANSTF Operational Plan; when warranted, utilize this information to update and amend the ANSTF Operational Plan

Objective 1.2: Evaluate the Ability of Statutory Authorities, Regulations, and Programs Necessary to Implement ANSTF Goals and Objectives

a) Identify gaps in statutory authorities, regulations, and programs necessary to meet ANSTF goals and objectives

b) Recommend revisions to statutory authorities, regulations, and programs when needed to meet ANSTF goals and objectives

Objective 1.3: Facilitate the Development and Continued Effectiveness of State and Interstate ANS Management Plans

a) Encourage plan development and provide technical drafting assistance to States without an ANS management plan

b) Encourage State and interstate ANS management plan review and revision every 5 years to reflect completed goals and objectives, new goals and objectives, new species and pathways, and management priorities

c) Seek opportunities to leverage funds to fully support implementation of ANSTF-approved State and interstate ANS management plans

d) Report on the number of State and interstate ANS management plans in place and under development

e) Review guidelines for the preparation of State and interstate ANS management plans

f) Annually summarize and report on State and interstate ANS management plan accomplishments and expenditures; utilize this information to compose a national assessment of ANS activity

Objective 1.4: Coordinate the Development and Implementation of ANSTF- Approved Species Control and Management Plans and Pathway Management Plans
   a)   Encourage the development of pathway management plans
   b)   Coordinate discussions on species and pathways for which plans may be needed; determine a lead entity for development of such plans
   c)   Distribute ANSTF-approved species management and control plans to State, tribal agencies, and relevant stakeholders to encourage State and tribal-level action
   d)   Monitor the development, evaluate the effectiveness of implementation, and report on the progress of ANS plan implementation
   e)   Report annually on the progress of existing ANSTF-approved species control and management plans
   f)   Support the implementation and long-term efforts of ANSTF approved plans

Objective 1.5: Cooperate with Nations That Have Neighboring Waters and Shared Pathways with the United States to Prevent, Detect, and Control ANS
   a)   Broaden involvement with international ANS activities and organizations
   b)   Continue and expand cooperation between ANSTF, Regional Panels, and foreign entities concerning the planning and implementation of prevention, monitoring, research, education, and control programs related to ANS that infest waters of the United States and neighboring nations

Goal 2: Prevention - Develop Strategies to Identify and Prevent the Establishment of New and Slow the Spread of Existing ANS in the Waters of the United States

Prevention is the first line of defense against ANS. Once a species becomes established, control efforts require significant and sustained resources. Since eradication may not be feasible, prevention is the most cost-effective means to avert the risk of harmful introductions. Investment in prevention avoids many of the long-term economic, environmental, and social costs associated with ANS. New species can arrive through many different ways, but most species that are considered to be invasive are a direct result of human activity. The movement of ANS may utilize pathways including, but not limited to, ballast water and hulls of ships, canals and waterways, aquaculture, the aquarium and pet trade, the bait industry, recreational activities, biological research, and habitat restoration projects.

Long-term success in prevention will reduce the rate of introductions, the rate of establishment, and the damage from additional ANS. One example of this success can be found in the Great Lakes. Beginning in 1970, one invader on average was recorded every 8 months in this region[22]; in 2006 more stringent measures were taken to regulate ballast water discharges, and since that time no new invasive species attributed to ballast water release and transoceanic shipping in general have been reported in the Great Lakes.[23] Additional successes in prevention will require Federal agency support and cooperation with State agencies and private organizations. Implementation of preventative measures may require broadening legislative mandates, strengthening the capacity of some departments, and refining or consolidating legislative and regulatory tools. The actions suggested below identify the most efficient way to reduce ANS risks by supporting authorities and programs that address intentional and unintentional introductions from all pathways. The joint ANSTF and NISC Prevention Committee via the Pathways Work Team has completed a report on major invasive species pathways (*Training and Implementation Guide for Pathway Definition, Risk Analysis and Risk Prioritization*) as well as a *Pathways Ranking Guide*. These documents will be used to guide ANSTF efforts in pathway management.

Objective 2.1: Take Steps to Interdict Specific Pathways by Developing and Implementing Guidance and Appropriate Measures

    a) Develop and implement risk mitigation measures to prevent introductions through priority pathways

    b) Evaluate risk mitigation measures to prevent introductions to ensure they are effective and environmentally sound

    c) Encourage State and Federal agencies to incorporate invasive species management into emergency response and contingency plans (e.g., wildfire management and spill response plans)

    d) Recommend amendments to the injurious wildlife provisions of the Lacey Act (18 U.S.C. § 42) to allow a proactive approach for preventing the establishment of new invasive species through the trade of live organisms

    e) Recommend amendments to legislation to ensure prevention of the establishment of nonnative aquatic plants though trade is addressed

    f) Encourage improved implementation and enforcement of the Lacey Act provisions on injurious wildlife and other regulations relevant to the transport, propagation, sale, collection, possession, importation, purchase, cultivation, distribution, and introduction of ANS

Objective 2.2: Facilitate Use of Science-Based Risk Assessment and Screening Procedures to Assess and Prioritize Pathways for the Introduction of ANS or Potential Species of Concern

    a) Identify authorities and regulations to carry out screening of specific pathways or species; recommend improvements based on identified gaps

    b) Support continued application of pathway ranking tools and related management guidance

    c) Develop and maintain a priority list for ANS pathways

Objective 2.3: Expand Training and Use of the Hazard Analysis and Critical Control Point (HACCP) Program into Work Conducted by Natural Resource Managers

Goal 3: Early Detection and Rapid Response - Identify and Respond to Aquatic Nuisance Species Within a Timely Manner Following Introduction in Order to Prevent Their Establishment and/or Spread

Despite the best preventive efforts, new ANS are certain to be introduced into waters of the United States. When a new species is introduced, the best strategy is early detection and rapid response (EDRR). This includes monitoring habitats to discover new species soon after introduction, reporting sightings of previously unknown species in an area, and working quickly to keep the species from becoming established and spreading. EDRR increases the likelihood that localized ANS populations will be found, contained, and eradicated before they become widely established. EDRR can slow range expansion of ANS, and avoid the need for costly long-term control efforts. Several States including Minnesota, Wisconsin, Indiana, and Virginia have successfully eradicated infestations preventing the establishment and spread of

ANS. Moreover, novel approaches to enhance early detection, including genetics-based methods, are being actively developed. These methods have strong potential to improve the ability to identify likely invaders and susceptible habitats.

Providing fast access to information on taxonomy, control methods, and subject matter experts can allow new and existing ANS to be readily recognized and managed. EDRR must be supported with emergency funding and guided by contingency plans coordinated by Federal, State, and local agencies as well as other entities, including tribes and non-governmental organizations. Goal 3 also supports several other needs: it evaluates prevention and control programs, provides information on invasion patterns and future management needs, and emphasizes the value of taxonomic expertise as an essential component of EDRR efforts. The actions suggested below identify environmentally sound methods that can prevent further spread and minimize harm to public interests. In addition to increasing ANS monitoring efforts, actions include the development of rapid response capabilities as well as research and education programs specifically related to the early detection of and rapid response to ANS.

Objective 3.1: Facilitate Surveys and Monitoring to Detect ANS

a) Assess existing early detection monitoring programs; identify gaps and recommend improvements for a more integrated approach
b) Develop model protocols for universal or common practices for early detection monitoring
c) Work with agencies and organizations to incorporate invasive species monitoring into existing survey work
d) Identify high-priority areas for targeted monitoring efforts; develop new early detection and surveillance programs as needed, including monitoring the impact of climate change on the distribution of ANS
e) Further develop, refine, and support genetic and emerging tools based on novel approaches to improve early detection of ANS

Objective 3.2: Make Taxonomic and Ecological Information and Expertise Readily Available

a) Update the ANSTF Expert Database on a regular basis
b) Identify gaps in available expertise
c) Support establishment of new taxonomic expertise to identify native and exotic species

Objective 3.3: Increase Public and Industry Involvement in Early Detection and Rapid Response Programs
    a) Increase public volunteer training opportunities utilizing existing programs and infrastructure
    b) Promote use of the ANS National Hotline and USGS Online Reporting Form to report sightings of non-native species
    c) Develop additional tools to encourage reporting of suspicious sightings

Objective 3.4: Facilitate Development of Rapid Response Contingency Plans
    a) Review and make accessible existing model rapid response plans for aquatic invasions
    b) Support development of additional model rapid response plans for aquatic invasions based on both taxonomic groups and jurisdictional authority where invasion occurred
    c) Identify authorities and regulations to carry out emergency response actions
    d) Support implementation of the National Incident Management System (NIMS) to prevent, protect against, recover from, and mitigate the effects of ANS incidents as mandated by Homeland Security Presidential Directive (HSPD)-5.[24]

Objective 3.5: Build Capacity to Respond Rapidly to Invasions
    a) Synthesize lessons learned from previous rapid response attempts to new ANS invasions and make it readily available
    b) Make species-specific control and management information readily available
    c) Develop a rapid response technical support network that that can provide resources and technical support in response to newly detected species
    d) Increase training opportunities of the NIMS protocol
    e) Increase number of mock-NIMS based rapid response exercises to identify additional steps needed for rapid response preparedness
    f) Explore opportunities and identify obstacles for establishing an emergency rapid response fund

Goal 4: Control and Management - Control Established Aquatic Nuisance Species When Feasible and When the Benefits of Managing the Established Species Outweigh the Costs of Removing Them with Respect to Harm to the Environment, the Economy, and Public Health

Once ANS are established, under most conditions complete eradication is usually not feasible. A more realistic approach for established populations is using control measures to slow the rate of range expansion and lessen the impacts to public interests. Management objectives may include eradication within an area, suppressing a population, limiting spread, and reducing impacts. Control measures may include mechanical, chemical, biological, and integrated pest management strategies. Adequate funding, public awareness, and management expertise are critical to success, particularly because ANS can span geographic and jurisdictional boundaries and do not recognize political boundaries or agency jurisdictions. Therefore, Federal and State agencies, Indian tribes, and private organizations should coordinate an ecosystem-level approach to managing ANS.

Multiple control and management tools are needed to assess, remove, and contain ANS populations as well as to guide management decisions. The actions suggested below seek to identify, improve, and execute these tools. Further, the Strategic Plan actions require inter-jurisdictional communication and regionally coordinated action through the continued development and implementation of control and management plans.

Objective 4.1: Support and Evaluate ANSTF-Approved Control and Management Plans

a) Identify gaps in control efforts and tools
b) When warranted, develop or broaden existing control methods and programs to achieve the target level of control

Objective 4.2: Increase Invasive Species Training for Natural Resource Managers and Leverage Participation

a) Increase the number of training workshops and total number of personnel and volunteers trained in control measures for ANS
b) Review and, as needed, develop ANS training materials used by natural resource managers

Objective 4.3: Evaluate the Benefits and Risks Associated with the Commercial Harvest of ANS as a Means of Control or Eradication
   a) Develop guidelines to assist States and tribes in determining when commercial harvest may be beneficial for control of ANS
   b) Develop guidelines to assist States and tribes in developing policies for new or existing commercial harvest programs for ANS
   c) Encourage long-term monitoring and evaluation of commercial harvest activities

Objective 4.4: Encourage an Integrated Pest Management (IPM)[25] Approach to Manage Existing ANS Populations

Goal 5: Restoration – Protect and Rehabilitate Native Species and Ecosystems by Conducting Habitat Restoration Efforts on Multiple Scales

Habitat restoration is an essential part of the control and management efforts used to guard against future invasions or to minimize harm to native ecological communities and other public interests. Restoration of the natural habitat should be addressed whenever the control or eradication of ANS is planned since habitat rehabilitation is often necessary to avoid the replacement of one invasive species with another, control flooding, or avoid other problems associated with the absence of biological organisms. Restoration activities may also include planting or stocking organisms or improving predator-prey relationships to attain food webs more similar to pre-invasion conditions. ANS can be transported by materials, equipment, vehicles, or personnel used to conduct restoration activities; accordingly all habitat restorations, even those not focused on ANS control, should call attention to actions that prevent establishment of invaders not yet present within the project site. Restoration efforts should make use of plant and animal species that are native to the particular habitat. One of the benefits of using native species includes their ability to thrive under the local conditions while being less likely to invade new habitats. Consequently, native species reduce maintenance costs and produce healthy natural communities, thus providing a practical and ecologically valuable option for restoration projects.

The actions suggested below focus on ANS concerns during habitat restoration efforts by targeting consideration of potential ANS during planning and implementation of restoration activities and encouraging post-restoration monitoring to ensure that any ANS introduced as a result of restoration are responded to in a rapid and efficient manner.

Objective 5.1: Restore Impacted Ecosystems

a) Identify and support agencies or programs that can assist in restoring areas impacted by ANS

b) Provide technical assistance on the species and methods to use in restoring native species, including means to enhance resilience against re-invasion, climate change, and other drivers of change

c) Compile, highlight, and share lessons learned for both restoration successes and failures within the United States

Objective 5.2: Address and Provide Technical Assistance for Invasive Species Management Before, During, and After Habitat Restoration Projects

a) Ensure that Federal land and water management field and guidance manuals consider ANS issues during the planning and development of habitat restoration projects

b) Review and make accessible existing restoration project standards to mitigate impacts of ANS during restoration activities. Develop new guidelines when warranted

c) Encourage application of adaptive management principles and assessment of treatment regimes to improve and sustain restoration efforts over time

d) Encourage the development of Hazard Analysis and Critical Control Point (HACCP) plans for all federally funded or authorized restoration projects

e) Support the development and expansion of markets that supply native plants and certified weed-free materials; encourage use of these materials by agencies and other organizations Encourage post-restoration monitoring for ANS by agencies and other organizations conducting habitat restoration or landscaping projects

f) Encourage restoration of areas following ANS eradication or control efforts

Goal 6: Education and Outreach - Increase Awareness Concerning the Threats of Aquatic Nuisance Species, Emphasizing the Impacts, Importance of Prevention and Containment, and Recommendations for Appropriate Domestic and International Actions

The lack of awareness concerning ANS impacts is one of the largest management obstacles. Few people understand the threat some nonindigenous species pose and how their actions might introduce them. Many ANS have been introduced through the actions of uninformed people; for example, disposing of bait, launching a boat, or stocking a private pond can each lead to the introduction of ANS if precautions are not taken. Further, the importation of organisms through trade has allowed species to spread by the receipt of unwanted organisms that hitchhike with the intentionally imported ones. Many policy makers, natural resource administrators, and private interest groups have facilitated the intentional introductions of species for certain economic or recreational purposes without understanding the effects these species may have on native species. These intentional and unintentional methods of introduction can be eliminated or curtailed by educating people about their potential to transfer ANS into new habitats.

Robust public awareness and action programs will help the public understand the impacts associated with invasive species so they can be partners in solving the problems. More importantly, people need to know what they can do to help prevent the introduction and spread of ANS in waters of the United States. ANSTF agency and *ex-officio* members, Regional Panels, States, tribes, and other entities have conducted workshops, created exhibits, pamphlets, information sheets, wallet identification cards, videos, websites, traveling displays, and other public education materials for distribution across the country. In recent years, many States have focused efforts on educating non-English speaking communities about ANS issues in general, and in respect to their culture. The actions suggested below focus on continuing to develop, review, and disseminate information to the general public as well as targeted user groups and businesses that may be potential vectors of ANS. Support and collaboration are necessary at many levels among and between Federal and State agencies, local governments, tribal entities, public and non-public sectors to successfully address ANS. As such, the actions below also focus on education of ANS threats and solutions for legislators and other decision makers.

Objective 6.1: Increase Understanding Among the General Public of the Problems and Impacts Associated with ANS and Actions That Can Be Taken to Prevent and Control ANS in Waters of the United States

    a)   Promote and expand new and existing national campaigns (e.g., Stop Aquatic Hitchhikers and Habitattitude™) with a proven track record of raising awareness and fostering behavior change among target audiences

    b)   Utilize the internet and social media as well as traditional media sources to disseminate information and promote awareness of ANS

    c)   Develop educational activities and products aimed at the general public regarding specific actions that can be taken to prevent, detect, and control ANS

    d)   Cooperate with media outlets to reach a broad range of the public with ANS messages

    e)   Participate in public affairs activities (e.g., conferences, shows, tournaments) to reach a broad range of the public with ANS messages

    f)   Raise the level of understanding and expertise on ANS worldwide by encouraging technical information exchange with other countries

    g)   Participate in international conferences and workshops

Objective 6.2: Disseminate ANS Outreach and Technical Guidance Materials to Target Audiences

    a)   Identify and prioritize targeted user groups and businesses that may be potential vectors of ANS.

    b)   Maintain and promote the ANSTF Recreational Guidelines

    c)   Leverage opportunities with relevant user groups and businesses to ensure awareness of the threats of ANS and reduce the risk of spread via emerging pathways

Objective 6.3: Promote the Use of Guidance Documents, Best Management Practices (Bmps),[26] and Other Outreach Materials Related to ANS

    a)   Encourage ANSTF member agencies, Regional Panels, and other stakeholders to submit guidance documents, BMPs, and other outreach materials to the ANSTF for review and endorsement

    b)   Identify gaps in outreach materials. Develop or update these materials when needed

c) Utilize the ANSTF website as a clearinghouse for guidance documents, BMPs, and other outreach materials endorsed by ANSTF

Objective 6.4: Promote Awareness of the ANSTF and Its Activities and Provide Educational Briefings on ANS Threats and Solutions and to Legislators and Other Decision Makers

a) Provide timely advice to the appropriate agencies concerning ANS that have been detected in waters of the United States, as well as waters of neighboring nations

b) Provide educational briefings and materials to Federal, State, tribal legislators and their staff members

c) Provide educational briefings and materials to Federal and State agency decision makers to build support for and incorporation of ANS programs into agency activities

d) Participate in and assist NISC with the national expansion of National Invasive Species Awareness Week

Goal 7: Research - Facilitate Research to Address Environmental, Economic, and Human Health Risks and Impacts Associated with Aquatic Nuisance Species

Information and research is needed to quantify and clarify the effects that ANS are having on native species and habitat as well as to socio-economics and human health. Although much research has been conducted for some aquatic invasive species, there are many species for which little is known. Increased knowledge of the biology, potential impacts, associated control methods, and interaction with climate change and other major drivers of change will allow for the most effective management of ANS. Research supports all facets of this Strategic Plan and is necessary to increase the effectiveness of prevention, detection, response, and control and management of invasive species. To help ensure that research addresses critical needs, the actions suggested below focus on coordination among Federal, State, and tribal governments; academia; and other participating entities. Economic research is also highlighted in this section. There is a lack of knowledge on a worldwide scale of the economic impacts of ANS. In many cases, it is the economic impacts that will be the driving force in effecting change in personal and business actions, management, and policy.

Many of the actions below encourage the continued development of risk analysis[27] tools to characterize the likelihood and severity of potential ANS to the environment, the economy, and human health and the means and methods to manage identified risks.[28] Science-based risk analysis is needed to evaluate invasive species before they reach the jurisdiction of the United States and to prioritize appropriate responses once they do. Risk analysis requires a methodology that integrates environmental, economic, social, and human health considerations. The principal role of ANSTF will be to provide guidance to these institutions on research, monitoring, and risk analysis needs and to provide feedback to researchers on the effectiveness of the management tools they develop.

Objective 7.1: Develop and Maintain a List of ANS Research Priorities; Communicate This List to the Scientific Community

Objective 7.2: Develop and Maintain Guidance Documents, Protocols, and Best Management Practices (BMPs) Related to ANS
   a)   Maintain and promote the Federal Aquatic Nuisance Species Research Risk Analysis Protocol[29]
   b)   Evaluate effectiveness and identify gaps in guidance documents and BMPs. Develop or update these documents when needed

Objective 7.3: Track the Progress of Research Activities Funded or Prioritized by the ANSTF
   a)   Utilize the research page on the ANSTF website as a clearinghouse for research activities funded or prioritized by the ANSTF
   b)   Share information on Federal invasive species grant opportunities and programs by linking this information from agency web pages to the ANSTF website

Objective 7.4: Support Development of Socio-Economic Research and Methods to Quantify the Economic Impact of ANS

Objective 7.5: Support Research on Interdiction Methods for Specific Pathways of ANS[30]
   a)   Facilitate the development of technologies and practices used for ballast water treatment and vessel hull fouling

b) Support development and implementation of fully effective ANS barriers between the Mississippi River and Great Lakes Basin and other infested natural waterbodies

Objective 7.6: Support Efforts to Identify Gaps and Expand Research Relevant to Control and Eradication Measures to Address ANS That Have Become Established in Waters of the United States

Objective 7.7: Encourage Research to Develop Species Invasion-Risk Forecast Tools

a) Develop and implement risk analyses and forecast models to evaluate invasiveness of high priority species, taking into account climate change and other drivers of change
b) Improve data collection at ports of entry so numbers and identification of species entering the United States through commerce in living organisms are available and accessible
c) Support development of a Risk Analysis Clearinghouse based on the outcome of species invasion-risk forecast analyses

Objective 7.8: Support Existing Databases and Global Database Networks So National and Worldwide Decision-Support Information for Invasive Species Management Is Accessible, Transparent, and Accurate

Goal 8: Funding - Coordinate Federal Agency Budgets to Support the Aquatic Nuisance Species Task Force's Priorities and Establish a Clear Process That Links State and Regional Needs in Their Areas of Responsibility

The ANSTF operates within a limited budget to conduct semiannual meetings and provides a fraction of the support needed to achieve goals identified by the Regional Panels and ANSTF-approved management plans. It is the cornerstone of the ANSTF to provide resources that will allow the States, Regional Panels, and tribes to implement programs that reflect the goals within the Strategic Plan. The actions suggested below focus on obtaining dedicated, long-term funding for the ANSTF by developing partnerships, and seeking opportunities to leverage funds within State and

Federal agencies, Indian tribes as well as public and private interests. The actions also encourage Federal agencies to continually review ANS priorities to find opportunities where agency authorities align with priority needs to create funding opportunities that can be met or communicated to the Office of Management and Budget.

Objective 8.1: Secure Dedicated, Long-Term Funding for the ANSTF Strategic Plan Actions
   a) Compose an annual report focused on ANS and use it as an opportunity to reach decision makers and other leaders on the need for proper policies and funding for ANS efforts
   b) Encourage Federal agencies to take ANSTF-approved Regional Panel recommendations into consideration as budgets are developed and to provide feedback
   c) Encourage Federal agencies to continually review regional priorities for opportunities where agency authorities align with priority needs in order to create funding opportunities that can either be met or communicated to the Office of Management and Budget
   d) Coordinate with NISC to ensure ANS priorities are recorded in the Interagency Invasive Species Performance Budget and communicated to the Office of Management and Budget
   e) Seek opportunities to leverage funds for ANS activities from Federal agencies and additional partners.

Objective 8.2: Optimize Use of Current Funding for ANS Activities by Engaging Potential Resources and Programs Within Federal Agencies and Additional Partners

Objective 8.3: Develop a List of ANS Funding Priorities
   a) Prioritize actions based on anticipated efficacy, threat level, and costs / benefits to natural resources
   b) Annually assess the funding needs of the Regional Panels as well as species-specific and State ANS management plans

APPENDIX 1: LIST OF ACRONYMS

ACOE	Army Corps of Engineers
AIS	Aquatic Invasive Species
ANS	Aquatic Nuisance Species
ANSTF	Aquatic Nuisance Species Task Force
APHIS	Animal and Plant Health Inspection Service
BLM	Bureau of Land Management
BMP	Best Management Practice
BOR	Bureau of Reclamation
DOS	Department of State
DOT	Department of Transportation
EDRR	Early Detection and Rapid Response
EPA	Environmental Protection Agency
FACA	Federal Advisory Committee Act
FICMNEW	Federal Interagency Committee for the Management of Nuisance and Exotic Weeds
HACCP	Hazard Analysis and Critical Control Point
HSDP	Homeland Security Presidential Directive
ICS	Incident Command System
IPM	Integrated Pest Management
ITAP	Federal Interagency Committee on Invasive Terrestrial Animals and Pathogens
MARAD	Maritime Administration
NANPCA	Nonindigenous Aquatic Nuisance Prevention and Control Act
NIMS	National Incident Management System
NISA	National Invasive Species Act
NISC	National Invasive Species Council
NOAA	National Oceanic and Atmospheric Administration
NPS	National Park Service
USCG	United States Coast Guard
USDA	United States Department of Agriculture
USFS	United States Forest Service
USFWS	United States Fish and Wildlife Service
USGS	United States Geological Survey

End Notes

[1] The term "vector" is continues to vary among agencies and organizations and is commonly confused with "pathway". The ANSTF defines a vector as the physical means or agent causing a species to translocate or spread (e.g., ship, car, waders).Pathway is defined as an activity or process through which a species may be transferred to a new location (e.g., shipping, animal trade, recreational activities).

[2] Wilcove DS, Rothstein D, Dubow J, Phillips A, Losos E. 1998. Quantifying threats to imperiled species in the United States. BioScience 48:607-615.

[3] Lawler JJ, Aukema JE, Grant, JB, Halpern BS, Kareiva P, Nelson CR, Ohleth K, Olden JD, Schlaepfer MA, Silliman BR, Zaradic P. 2006. Conservation science: a 20-year report card. Frontiers in Ecology and the Environment 4: 473-480.

[4] Zambrano L, Martinez-Meyer E, Menezes N, Peterson AT. 2006. Invasive potential of common carp (Cyprinus carpio) and Nile tilapia (Oreochromis niloticus) in American freshwater systems. Canadian Journal of Fisheries and Aquatic Sciences 63: 1903–1910.

[5] United States Department of Agriculture APHIS Fact Sheet. Invasive Species. October. 1999.

[6] Pimentel D, Zuniga R, Morrison D. 2005. Update on the environmental and economic costs associated with alien-invasive species in the United States. Ecological Economics 52:273–288.

[7] Simpson A. 2004. The Global Invasive Species Information Network: What's in it for you? BioScience 54: 613-614.

[8] Army, 2002. Economic Impacts of Zebra Mussel Infestation. http://www.wes.army. mil / el /zebra/zmis/zmis/zmishelp/economic_ impacts_of_zebra_mussel_infestation.htm (Accessed April 1, 2012).

[9] Center TD, Frank JH, Dray FA, 1997. Biological control. In: Simberloff D, Schmitz DC, Brown TC. (Eds.), Strangers in Paradise. Island Press, Washington, DC, pp. 245– 266.

[10] Walters LJ, Brown KR, Stam WT, and Olsen JL. 2006. Ecommerce and Caulerpa: unregulated dispersal of invasive species. Frontiers in Ecology and the Environment 4: 75–79.

[11] For the purposes of this document the term "Asian carps" refers to four species: black carp (Mylopharyngodon piceus), bighead carp (Hypophthalmichthys nobilis), grass carp (Ctenopharyngodon idella), and silver carp (H. molitrix).

[12] 2012 Asian Carp Control Strategy Framework. Asian Carp Regional Coordinating Committee. February 2012. http://asiancarp.us/documents/2012Framework.pdf. Accessed May 3, 2012.

[13] Ruiz GM, Rawlings TK, Dobbs FC, Drake LA, Mullady T, Huq A, Colwell RR. 2000. Global spread of microorganisms by ships. Nature 408:49–50.

[14] Boesch DF, Anderson DM, Horner RA, Shumway SE, Tester PA, Whitledge TE. 1997. Harmful Algal Blooms in Coastal Waters: Options for Prevention, Control, and Mitigation. NOAA Coastal Ocean Program Decision Analysis Series No. 10. NOAA Coastal Ocean Office, Silver Spring, MD. 46pp. + appendix.

[15] Cohen, A.N. 2010. Non-native Bacterial and Viral Pathogens in Ballast Water: Potential for Impacts to ESA-listed Species under NOAA's Jurisdiction. A report prepared for the National Oceanic and Atmospheric Administration, National Marine Fisheries Service, Endangered Species Division, Silver Spring, MD. Center for Research on Aquatic Bioinvasions (CRAB), Richmond, CA

[16] Carlton JT. 1996. Marine bioinvasions: The alternation of marine ecosystems by nonindigenous species. Oceanography. 9: 36-43.

[17] The term "waters of the United States" is defined by the Clean Water Act 40 CFR 230.3(s)

[18] Federal agencies specified within the Act include the U.S. Fish and Wildlife Service, National Oceanic and Atmospheric Administration, Environmental Protection Agency, United States Coast Guard, Army Corps of Engineers, United States Department of Agriculture.

[19] Organizations specified within the Act include the Great Lakes Commission, Lake Champlain Basin Program, Chesapeake Bay Program, and San Francisco Bay-Delta Estuary Program.

[20] Two members co-represent the Native American Fish and Wildlife Society

[21] Section 1202(k) (2) of NANPCA, requires the ANSTF to submit, on an annual basis, a report to Congress focusing on progress in carrying out the provisions of the Act. Under the Act, the Regional Panels are required to submit an annual report to the ANSTF describing activities in their regions related to ANS prevention, research and control activities. Additionally, contracts for funding the panels require an annual report.

[22] Ricciardi A. 2001. Facilitative interactions among aquatic invaders: is an "invasional meltdown" occurring in the Great Lakes? Can Canadian Journal of Fisheries and Aquatic Sciences 58: 2513–2525.

[23] Bailey SA, Deneau MG, Jean L, Wiley CJ, Leung B, MacIsaac HJ. 2011. Evaluating efficacy of an environmental policy to prevent biological invasions. Environmental Science and Technology 45: 2554–2561.

[24] Homeland Security Presidential Directive 5 (HSPD-5) required all Federal agencies and departments to adopt the National Incident Management System (NIMS) to coordinate emergency preparedness and incident management and response among the public and private sectors.

[25] Integrated pest management (IPM) is a broad based ecological approach that utilizes a range of practices to maximize control of a species. In IPM, one attempts to prevent introduction, to observe patterns of spread, and control as necessary by the most economical means, and with the least possible hazard to people, property, and the environment.

[26] The phrase "Best Management Practice" (BMP) was originally used in the U.S. Clean Water Act and associated Federal regulations to refer to procedures used for industrial wastewater control and stormwater management. Although the term was defined by the Environmental Protection Agency as a regulatory tool used to implement Federal wastewater permit regulations, it is now commonly used in the language of environmental management. The Aquatic Nuisance Species Task Force uses the term BMP in the Strategic and Operational Plans to describe any method or technique found to be the most effective and practical means in achieving an objective while making optimum use of resources.

[27] The term "Risk analysis" in this document includes both risk assessment and risk management. Risk assessment measures the likelihood of an event occurring and the severity of negative impacts from such an event. Risk management is the process of identifying, evaluating, selecting, and implementing actions to reduce risk and includes more subjective elements including risk prioritization, risk tolerance, and associated decisions that weigh the benefits and costs of risk minimization options.

[28] Additional information on ANS risk analysis methodology can be found within 1) National Research Council. 1983. Risk Assessment in the Federal Government: Managing the Process. National Academy Press, Washington DC. 2) National Research Council. 1993. Issues in Risk Assessment. National Academy Press, Washington DC. 3) Stern, P.C. and H.V. Fineberg (eds). 1996. Understanding Risk: Informing Decisions in a Democratic Society. National Academy Press, Washington DC. 4) National Research Council. 2009. Science and Decisions: Advancing Risk Assessment. National Academy Press, Washington DC.

[29] The ANSTF developed a research protocol as is required by the Nonindigenous Aquatic Nuisance Prevention and Control Act of 1990 (NANPCA, 101, 104 STAT. 4671, 16 U.S.C. 4701-4741), as amended by the National Invasive Species Act, 1996. Section 1202(f) (2) of NANPCA directs the ANSTF to establish a protocol "to ensure that research activities carried out under [NANPCA] do not result in the introduction of aquatic nuisance species to waters of the United States."

[30] The action items listed for Objective 7.5 are included in this document as they are mandated by the Nonindigenous Aquatic Nuisance Prevention and Control Act (NANPCA). Additional actions taken to support this objective will be included in the ANSTF Operational Plan.

INDEX